Nanoscale Science and Engineering Forum 2019

Held at the 2019 AIChE Annual Meeting

Orlando, Florida, USA
10 - 15 November 2019

ISBN: 978-1-7138-0543-4

Nanoscale Science and Engineering Forum 2019

Held at the 2019 AIChE Annual Meeting

Orlando, Florida, USA
10 - 15 November 2019

AIChE
The Global Home of Chemical Engineers

Printed from e-media with permission by:

Curran Associates, Inc.
57 Morehouse Lane
Red Hook, NY 12571

Some format issues inherent in the e-media version may also appear in this print version.

Printed with permission by Curran Associates, Inc. (2020)

For permission requests, please contact AIChE
at the address below.

AIChE
120 Wall Street, FL 23
New York, NY 10005-4020

Phone: (800) 242-4363
Fax: (203) 775-5177

www.aiche.org

Additional copies of this publication are available from:

Curran Associates, Inc.
57 Morehouse Lane
Red Hook, NY 12571 USA
Phone: 845-758-0400
Fax: 845-758-2633
Email: curran@proceedings.com
Web: www.proceedings.com

TABLE OF CONTENTS

580133 Progelator Peptides and Peptide-Polymer Conjugates for Treating Myocardial Infarction

Injectable biopolymer hydrogels have gained increasing attention for clinical translation as engineered biomedical scaffolds to promote cardiac function and prevent negative left ventricular (LV) remodeling post-MI). However, the majority of hydrogels tested are not candidates for minimally invasive catheter delivery because of excess material viscosity, rapid gelation times, and concerns regarding hemocompatibility and potential for embolism. As a result, clinical translation of most injectable biomaterials for the heart has been hindered. Solutions are needed for this broad class of promising biomaterial.

Here, we describe novel progelator materials formulated in some cases as sterically constrained cyclic peptides to prevent self-assembly and provide a free flowing solution for low resistance injection. Importantly through our design, we provide the first demonstration of self-assembling peptides (SAPs), and related polymer-peptide conjugate scaffolds, amenable to cardiac catheter injection. Ulitmate linearization by MI-associated enzymes induces rapid self-assembly into rehealable hydrogels. The versatility of our platform is shown through the functionalization of two different SAP sequences, which exhibit disparate self-assembly mechanisms, and yet form progelators with identical responsiveness. Hemocompatibility analyses and in vivo application in a rat ischemia repurfusion model provide evidence that our simple synthetic modifications do not induce toxicity nor alter the capacity to self-assemble within a biological environment. Finally, we demonstrate that labeling of our progelators with a small molecule dye (rhodamine) does not interfere with self-assembly. Thus we envision that a chemically complex hydrogel can be generated in vivo through simple mixture of different progelators, each bearing a small molecule drug, tag, or reactive moiety. The work presented sets the stage for structurally dynamic biomaterials for therapeutic hydrogel delivery to the heart for the prevention of negative LV remodeling.

577695 Light-Responsive Fluorescently-Labeled Cyclic Peptides for Selective and Reversible Labeling of Therapeutic Cells

Abstract. One of the major challenges in mainstream cell-sorting technology is the removal of the affinity tags utilized for labeling the target cell population. Monoclonal antibodies targeting cell surface receptors, while highly selective, feature very high binding strength, which prevents their removal from the purified cells. To overcome this issue, we have developed a method for selective and reversible cell labeling, based on photo-switchable peptides that bind cell surface markers in an ON/OFF manner upon exposure to light. These peptides comprise a marker-targeting amino acid segment cyclized by an azobenzene linker, and tagged with a fluorescent dye. Exposure to light triggers *cis/trans* isomerization of the azobenzene linker, causing the peptide to switch reversibly between a binding and non-binding configuration. In this work, we have developed a fluorescein-labeled cyclic peptide specific for the surface marker VCAM1. Starting from a known VCAM1 binding linear peptide, an ensemble of azobenzene-cyclic variants were designed and optimized to ensure high binding selectivity and ON/OFF-type light-controlled binding. To this end, we combined *in silico* design using molecular dynamics and docking, with experimental sequence optimization by spectroscopy and surface plasmon resonance (SPR). The peptide lead $cyclo_{AZOB}$[G-VHAKQHRN-K] in particular displayed efficient photo-switching and ability to bind and release VCAM1 in solution with high yield upon exposure to light. Fluorescein-$cyclo_{AZOB}$[G-VHAKQHRN-K] was then utilized to label human umbilical vein endothelial cells (HUVECs), where VCAM1 expression was induced with glucose, in a light-dependent manner. The peptide, in fact, bound HUVECs in *trans* configuration, while showing negligible binding in *cis*, thereby demonstrating the potential of the proposed method for identifying peptides for cell labeling and purification.

569002 Investigating Sequence Effects on the Biodegradation of Peptoid Substrates

The structure and function of biological molecules arise from the precise sequence of building blocks responsible for encoding their properties. Nature's strategy for structure-property regulation has long been an inspiration in the polymer chemistry field, aiming to combine sequence-controlled synthesis methods with the customizable and robust qualities of abiotic backbones. Sequence control is especially applicable to biomaterials, where tunable properties are necessary to match the dynamic criteria encountered *in vivo*. Implementing sequence control in synthetic platforms expands the range of properties available compared to purely biological backbones and affords adaptability through sequence, rather than compositional changes. Peptoids, or N-substituted glycines, are peptidomimetics with sidechains affixed to the amide backbone nitrogen, rather than the α-carbon. Peptoids can be synthesized with full sequence definition using the submonomer technique, which affords incorporation of any sidechain available as a primary amine. Thus, peptoids provide a versatile platform for advancing biomaterials, but their functional diversity necessitates thorough characterization of peptoid properties in relevance to sidechain chemistry and sequence effects. Here, we investigated sequence effects on the biodegradation of peptoid-peptide hybrids for potential use in synthetic biomaterials, namely by enzymatic and oxidative mechanisms. First, we examined the role of peptoid substitutions on the active site of matrix metalloproteinase (MMP)-degradable substrates by synthesizing a control peptide (VPLS↓LYSG, where ↓ indicates cleavage site) and a peptide-peptoid hybrid with a peptoid substitution at the P3 subsite (N-[2-(1-pyrrolidinyl)ethyl]glycine substituted for proline). We found that collagenase was able to hydrolyze this sequence completely at kinetic rates similar to the control substrate. Next, we examined the oxidative susceptibility of peptide-peptoid hybrids by synthesizing molecules composed of N-(2-methoxyethyl)glycine and proline residues in various sequence patterns. These substrates were incubated with hydrogen peroxide, and their degradation was monitored over time via liquid chromatography-mass spectrometry (LC-MS). We found that the peptoid backbone is susceptible to backbone oxidative degradation, similar to proline. Ongoing work will continue to expand the library of substrates under investigation to probe the design rules dictating enzyme-substrate selectivity and oxidative susceptibility. Ultimately, these results indicate that sequence-defined N-substitutions can be used to engineer substrates with tunable degradative properties for implementation in stimuli-responsive biomaterials. Comprehensive investigation of their degradative properties could allow for programmable degradation profiles and selective substrates amenable to both enzymatic and oxidative response mechanisms.

576508 Self-Assembled Tri-Peptides As Water-Responsive Materials

Water-responsive (WR) materials that respond to relative humidity alternations by changing their shapes are receiving growing interest due to their potential applications, including the development of evaporation energy harvesting technique and WR artificial muscles. Materials with WR properties are widely observed in natural systems, such as pine cones, spider silk, and wheat awns. While WR mechanisms of these natural materials are still poorly understood, they share many common features, including hierarchical order, nanoscale pores, and ordered/disordered domains in their structures that are often composed of peptides and glycans. Inspired by nature, we crystalize short peptides into high performance WR materials. We demonstrate that these self-assembled short peptide crystals exhibit high WR strain (>25%), structural stability, mechanical stiffness (>2GPa), and high WR energy density (> 80kJ/m^3).

572095 A Kinetic Path to Hierarchically Organized Soft Materials from Self-Assembling Discotics

Introduction: Proteins and other structural building blocks in nature traverse complex free energy landscapes to create hierarchically organized materials. For instance, collagen is a ubiquitous biopolymer which underlies structure and function in tissues. Many variants of collagen begin with triple-helical folding and further self-assemble into fibrils and higher order bundles. These end-state structures impart biological functions critical to the extracellular matrix (ECM) that surrounds cells, offering binding and receptor sites for various endogenous proteins and other biomolecules. Of note, the self-association pattern of the collagen triple-helix has been shown to be dependent on tissue environment and collagen protein type, generating a diversity of functional forms ranging from anchoring fibrils to vitreous gelators, guided by supramolecular cues directing the assembly pathway. One particular outcome of interest is the large laterally bundled network components that comprise the ECM, which facilitate cell binding, differentiation, and growth. We developed a model discostic amphiphile gelator based on the benzene tricarboxamide (BTA) supramolecular motif, which undergoes pathway-dependent structural maturation into triple-helices similar to that of collagen and collagen-like proteins, under varying rate of pH control as the guiding stimulus.

Methods: Isopeptides were synthesized with solid phase methods and conjugated to trimesoyl trichloride to yield isopeptide-bearing benzene tricarboxamide molecules, of which the three-arm species was isolated using HPLC. Hydrogels were then prepared via acidification of the samples mediated by either hydrochloric acid (fast) or glucono-d-lactone (slow), and divalent calcium ions were contributed by either calcium chloride (fast dissociation) or calcium carbonate (slow dissociation). Gelation kinetics and viscoelastic properties of these hydrogels were investigated by oscillating rheometry. Circular dichroism spectroscopy was used to track the emergence of distinctive chiral signatures arising in the materials over time due to intrinsic chiral features ascribed to self-assembled individual stacks and triple helical fibrils. Transmission and scanning electron microscopy were employed to study the morphology of the self-assembled structures comprising the hydrogel. Polarized optical microscopy was used to study potential birefringent character imparted by the liquid crystalline nature of the hydrogels.

Results: Through slow pH change, complex hierarchical assemblies result, characterized by mesoscale elongated rod-like bundles embedded in a percolated mesh of narrow triple-helical fibers. In contrast, in the case of rapid pH change, the formation of such mesoscale superstructures is very minimal or non-existent. Furthermore, the presence of divalent cations in the system proved impactful, with markedly different outcomes arising in the cases of quickly vs slowly dissociating salts vs no divalent cation source, suggesting another dimension of dynamicity exhibited in this system. These cations were interpreted to play a critical role in mediating electrostatic bridging / charge screening that enables lateral bundling as well as one-dimensional growth, to different extents. All rheological hydrogels arising from these catalyzed self-assembly processes bear shear-thinning, self-healing features that would facilitate ease of passage through a syringe needle and subsequent restoration of gel state upon cessation of shear. The storage modulus of these hydrogels was furthermore dependent on the assembly pathway, with emergent hierarchical organization and lateral bundling conferring increased mechanical stiffness.

Conclusions: The broad range of heterogeneity in materials prepared from a single self-assembling gelator implies a measure of thermodynamic pathway control imparted by the gelation stimulus selection, and also suggests foreseeable merit in the pursuit of a collagen-mimetic biomaterial constructed from BTAs as a base unit. Their shear-thinning self-healing nature enables uses as an injectable ECM-mimicking material that could have application in regenerative engineering or cell-based therapy. Finally, modification of BTA substituents enables a catalogue of material building blocks analogous to the collagen family of proteins, where selection of one or more BTA type could give rise to a distinct material outcome.

576640 Nanoparticles Self-Assembly for the Preparation of Bioinspired Materials with Stimuli-Responsive Color Changing Ability

Octopuses and cuttlefish possess some of the most amazing camouflage abilities of the entire animal kingdom. The core of the camouflage mechanism lies in chromatophores, which are special cells with the ability to displace pigments, located in small pockets in the middle of the cell, over a much larger surface, thus enabling the overall appearance of the cell to change from white or transparent to the specific color of the pigment. Inspired by this mechanism, we have designed color changing polymeric materials exploiting the self-assembly of responsive colloids. The design of the material, unique in its kind, is based on the incorporation of water droplets, generated by microfluidics, in which magnetic nanoclusters are well dispersed, and impart a brown color to the droplets. These droplets play the same role as chromatophores in the octopus's skin. However, upon exposure to a magnetic field gradient, the clusters rapidly self-organize, align in the direction of the field or accumulate in a small region of the droplet, towards the direction of the strongest magnetic field gradient, thus mimicking the pigment dislocation in the chromatophores. In this manner, the overall color of the droplets changes from dark brown to almost transparent, and leads to macroscopic change in the overall color of the material. To the best of our knowledge, this is the first example of a material with magnetically responsive color change mimicking the pigment dislocation effect.

577641 Protein-Engineered Hybrid Nanomaterials with Complex Morphologies

In nature, biological processes are regulated by a variety of protein domains that specifically interact. Strategies that exploit such interactions between proteins and inorganics have emerged as a biomimetic strategy to construct hybrid nanomaterials. For example, peptides, proteins, or protein nano-assemblies that are involved in natural or synthetic biomineralization have been engineered and used for hybrid material fabrication. However, controls over the dimension or morphology of resulting hybrid structures have been challenged by limited availability of robust biotemplates or complicated material assembly pathways. In this work, we present an approach to complex-nanostructured hybrid materials that are created by *de novo* design of versatile nanoscale protein-cage templates. Recombinant protein building blocks were created by modularly and genetically integrating multiple types of protein domains that self-assemble or interact with non-biological components such as synthetic polymers or inorganic materials. The programmed self-assembling interactions between the building blocks directed assembly of protein nanocages that self-templated formation of hybrid nanostructures. In this protein-engineered hybrid nanomaterials, polymeric or inorganic nanoparticles were inserted and placed inside the protein nanocages, resulting in jingle bell or pomegranate-like morphologies. These hybrid materials present a unique structural feature that nanoscale confined space is provided between protein layers and inserted materials. We demonstrate material morphology tunability, mechanisms that describe complicated interactions between proteins and synthetic components, and potential applications using optically or electronically active hybrid materials.

573429 Carbon Nanotube Carpet Growth on Nontraditional Substrates Using an Industrial Gaseous Waste As a Feedstock

Carbon Nanotube Carpet Growth on Nontraditional Substrates using an Industrial Gaseous Waste as a Feedstock

Aaron Porter, Xu Li, Placidus Amama

Tim Taylor Department of Chemical Engineering, Kansas State University, Manhattan KS, 66506

There is increasing interest in achieving efficient growth of vertically aligned, high-quality carbon nanotube arrays (CNT carpets). This is due to their suitability in a wide range of applications such as energy storage, thermally conductive materials, catalysis, and sensing devices. The preferred method for CNT synthesis is chemical vapor deposition (CVD) due to its simplicity and potential of achieving mass production of CNTs with tailored properties. Typically, the catalytic CVD process involves CNT carpet growth from catalyst particles that are transition metals (Fe, Co, or Ni). These particles are supported on a thin film of alumina or silica. However, some of the applications mentioned above require CNT arrays to be grown directly on conductive, temperature-sensitive, easily removable substrates or non-alumina supports. In this study, we explore growth of CNT carpets from a variety of substrates with different properties, considering composition, phase, basicity, and surface properties. We will test growth with a conventional feedstock (ethylene), and our newly developed feedstock (industrial waste gas from Fischer-Tropsch synthesis). By these means we aim to establish interrelationships between substrate properties and CNT carpet growth behavior.

567863 Production of CNT Yarns from Methane Gas for Use As Filaments in Incandescent Bulbs: Effect of Hydrogen Co-Feeding on the Properties of As-Spun CNT Yarns

In spite of the excellent electrical, mechanical and optical properties of carbon nanotubes (CNTs), their application in electronics devices have been limited. Therefore, organization of CNTs into their macroscopic sizes while maintaining the quality of the individual CNTs could be instrumental to paving the way for further application of CNTs, e.g. in incandescent bulb. Application of CNTs as filament to replace tungsten in incandescent bulbs requires the assembly of CNTs into macroscopic strings called yarns. Tungsten as a filament in incandescent bulbs is currently being phased out and undergoing outright ban in many countries owing to its very low efficiency (only about 5% of the energy input into incandescent bulbs with tungsten as filament is used for lightening and the rest is dissipated as heat and resultant emissions) [1]. However, the use of CNT yarns as filament presents an option for improving the environmental and economical inefficiency of tungsten based incandescent bulbs. In addition, co-feeding of hydrogen during the production of CNTs has been reported to be detrimental to the morphology, electrical conductivity and mechanical properties of synthesized CNT yarns [2].

Spontaneous assembly of single walled, double walled and even mixed CNTs during synthesis due to Van der Vaal forces [3, 4] is enhanced in this study, where carbon nanotubes were synthesized using methane as carbon source and ferrocene as catalyst and spun into yarns in a vertical Swirled Floating Catalyst Chemical Vapour Deposition (SFCCVD) reactor. Furthermore, investigation of the effect of hydrogen co-feeding on length, compactness and number of walls of the CNTs produced from saturated hydrocarbon feedstock was carried out.

Experiments are on-going, however, it is expected that the results of the investigation will shed light on the relationship that exists between the mechanical property and electrical property of the as-produced CNTs as will be obtained from Raman spectroscopy and X-ray diffraction. Furthermore, the study will provide information on the effect of hydrogen co-feeding on the morphology, crystallinity, mechanical property and electrical conductivity of the as-produced yarns. The results could form a basis to arguably compare the obtained results with literature such as the report of Tsentalovich et al [5]. At the moment, results from our preliminary investigations indicate the predominance of multi-walled CNTs during the production.

References

[1] MacIsaac, D., Kanner, G., & Anderson, G. (1999). Basic physics of the incandescent lamp (lightbulb). The physics teacher, 37(9), 520-525.

[2] Ma, Y., Dichiara, A. B., He, D., Zimmer, L., & Bai, J. (2016). Control of product nature and morphology by adjusting the hydrogen content in a continuous chemical vapor deposition process for carbon nanotube synthesis. Carbon, 107, 171-179.

[3] Li, Y., Zhang, X. B., Tao, X. Y., Xu, J. M., Huang, W. Z., Luo, J. H., ... & Geise, H. J. (2005). Mass production of high-quality multi-walled carbon nanotube bundles on a Ni/Mo/MgO catalyst. Carbon, 43(2), 295-301.

[4] Colomer, J. F., Henrard, L., Van Tendeloo, G., Lucas, A., & Lambin, P. (2004). Study of the packing of double-walled carbon nanotubes into bundles by transmission electron microscopy and electron diffraction. Journal of Materials Chemistry, 14(4), 603-606.

[5] Tsentalovich, D. E., Headrick, R. J., Mirri, F., Hao, J., Behabtu, N., Young, C. C., & Pasquali, M. (2017). Influence of carbon nanotube characteristics on macroscopic fiber properties. ACS applied materials & interfaces, 9(41), 36189-36198.

575044 Brief Review of Thermal Graphitization of Coals and Carbons

Brief Review of Thermal Graphitization of Coals and Carbons

Evan J. Granite

United States Department of Energy

National Energy Technology Laboratory

Pittsburgh, PA 15236-0940

evan.granite@netl.doe.gov

Abstract

The United States has the world's largest coal reserves. Graphite and graphite-like materials can be interesting products from coal and other low-cost carbons. Coal forms from plant matter over geological time periods; i.e., millions to hundreds of millions of years. As the plant matter decomposes, the material becomes more ordered (coalified or graphitized) under the influence of time, temperature and pressure. Lower ranks of coal such as lignite are younger and contain less fixed carbon, visible plant material, more volatile matter and less ordered carbon. In contrast, high rank coals (anthracite) develop structures that resemble graphite. With the formation of high rank coals in nature as a starting point, this talk will briefly review thermal processes for the graphitization of coals and other carbon precursors.

Disclaimer

576740 Multifunctional Hydrogel-Coated Gold Nanoshells for Protein Biomarker Quantification

There has been growing interest in using biomarkers expressed in readily available patient samples, such as saliva, tears, urine, or serum, to develop minimally-invasive tests for disease screening and diagnosis. While many different approaches have been explored, differential sensors exploiting arrays of semi-selective recognitive elements are a promising strategy for extending current methods to multiple marker quantification without the need for the development of a new receptor for each biomarker. However, translating differential sensing approaches to protein recognition is challenged by the similarity of the biomarkers' structures and properties to those of the other proteins present in the sample. In order address this challenge, we explore the capabilities of hydrogel network formation on the surface of gold nanoshells (AuNS) as a semi-selective recognition element for the quantification of protein biomarkers based on differences in polymer functionalization using shifts in the maximum localized surface plasmon resonance (LSPR) wavelength.

AuNS were prepared by seeded growth of colloidal gold on the surface of aminated silica nanoparticles followed by encapsulation with poly(maleic anhydride-*alt*-1-octadecene)-g-poly(ethylene glycol) methacrylate (PMAO-g-PEGMA) graft copolymer to improve particle stability. Hydrogel shells were formed by incorporating the AuNS into the aqueous phase of an inverse emulsion copolymerization of methacrylic acid with acrylamide or n-isopropylacrylamide, crosslinked with N,N'-methylenebis(acrylamide). Transmission electron microscopy was used to determine nanoshell size distribution and UV-Vis spectroscopy was used to measure the shifts in the wavelength of the LSPR. These polymer-coated AuNS can serve as a semi-selective sensor for protein adsorption, which will produce a red shift in the LSPR, and can be modified by incorporating comonomers with selective functionalities or by post-synthesis modification through carbodiimide coupling in order to improve the selectivity of the sensor for certain biomarkers.

Work supported by National Institutes of Health Grant R01-EB022025

574741 Surface Effects of Ultrasmall Nanoparticle on Cellular Uptake, Proliferation, and Multipotency of Neural Stem Cells

Ultrasmall iron oxide nanoparticles (USIONPs) have been recently developed as novel labelling probes as positive contrast agents for magnetic resonance imaging (MRI). However, their use in stem cell tracking have not been reported. Here, we studied the effects of USIONP surface coatings on proliferation, cellular uptake, and multipontency of established neural stem cells (NSCs) and primary NSCs. USIONPs were functionalized with gluconic acid (GA), tannic acid (TA), and hyaluronic acid (HA) to label NSCs. All functionalized USIONPs were characterized as T1 contrast agents via relaxivity measurements. Nanoparticles functionalized with TA and HA-coating promoted NSC proliferation and enhanced cellular uptake in a dose-dependent manner. Furthermore, HA-coating showed enhanced cell proliferation and cellular uptake in primary NSCs depending on the HA molecular weight. Stem cell characteristics were well-maintained, verified by neurosphere formation and gene expression levels of stemness and differentiation markers. Collectively, we demonstrated that NSC proliferation, cellular uptake, and multipotency can be enhanced using various surface coating strategy of USIONPs.

575596 Quantum Capacitance Based Amplified Graphene Phononics for Studying Neurodegenerative Diseases

Amyotrophic lateral sclerosis (ALS) is the most common adult-onset motor neuron disease, characterized by a rapid loss of upper and lower motor-neurons resulting in patient death from respiratory failure within 3-5 years of initial symptoms onset. Although at least 30 genes of major effect have been reported, the pathobiology of ALS is not well understood. Compounding this is the lack of a reliable laboratory test which can accurately diagnose this rapidly deteriorating disease. Herein, we report on graphene's phonon vibration-energies as a sensitive measure of the composite dipole moment of the components of the interfaced cerebrospinal fluid (CSF) to specifically identify patients with ALS disease. The second-order overtone of in-plane phonon vibration energy (2D) of graphene shifts by 3.2 ± 0.5 cm^{-1} for all ALS patients studied in this work. Further, the amount of n-doping induced shift in phonon energy of graphene, interfaced with CSF, is specific to the investigated neurodegenerative disease (ALS, Multiple Sclerosis and Motor Neuron Disease). By removing a severe roadblock in disease detection, this technology can be applied to study diagnostic biomarkers for researchers developing therapeutics and clinicians initiating treatments for neurodegenerative diseases.

567943 Rare Earth Element Doped Nanocrystals for Correlative Cathodoluminescence Electron Microscopy Bioimaging (CCLEM)

Correlative imaging has enabled some of the biggest scientific discoveries in the past. It has become a powerful analytical tool for unravelling structure-function relationships. However, correlating optical and electron microscopy is often challenging due to the resolution mismatch and different sample preparation requirements.[1] Interestingly, accelerated electrons can also generate luminescence signal, creating an effect called cathodoluminescence (CL)[2, 3]. Correlative cathodoluminescence electron microscopy (CCLEM) bioimaging correlates cathodoluminescence and electron microscopy images[4]. Although luminescence excitation by an electron beam enables sub-diffraction imaging, structural damage to the sample by high energy electrons has been identified as a potential obstacle.

Here, we present rare earth (RE) element doped nanocrystals as bright and stable probes for high-quality CCLEM. We report on the performance of RE-doped nanocrystals in CCLEM compared to fluorophores and quantum dots commonly used as molecular labels in biology. The influence of the electron beam parameters on image quality and resolution will be presented along with the first experimental demonstration of correlative cathodoluminescence back-scattering electron microscopy on focused ion beam (FIB)-sectioned biological samples[4]. We also demonstrate how exogenous nanoparticles can be easily distinguished from naturally occurring cellular features based on the characteristic optical emission. Taken together, we demonstrate the potential of the CCLEM approach for the acquisition of high-quality luminescence and electron microscopy images with nanometric resolution of biological samples.

1. Hell, S. W.; Sahl, S. J.; Bates, M.; Zhuang, X.; Heintzmann, R.; Booth, M. J.; Bewersdorf, J.; Shtengel, G.; Hess, H.; Tinnefeld, P.; Honigmann, A.; Jakobs, S.; Testa, I.; Cognet, L.; Lounis, B.; Ewers, H.; Davis, S. J.; Eggeling, C.; Klenerman, D.; Willig, K. I.; Vicidomini, G.; Castello, M.; Diaspro, A.; Cordes, T. *Journal of Physics D: Applied Physics* **2015,** 48, (44), 443001.
2. Mahfoud, Z.; Dijksman, A. T.; Javaux, C.; Bassoul, P.; Baudrion, A.-L.; Plain, J.; Dubertret, B.; Kociak, M. *The Journal of Physical Chemistry Letters* **2013,** 4, (23), 4090-4094.
3. Atre, A. C.; Brenny, B. J. M.; Coenen, T.; García-Etxarri, A.; Polman, A.; Dionne, J. A. *Nature Nanotechnology* **2015,** 10, 429.
4. Keevend, K.; Stiefel, M.; Neuer, A. L.; Matter, M. T.; Neels, A.; Bertazzo, S.; Herrmann, I. K. *Nanoscale* **2017,** 9, (13), 4383-4387.

568345 Effect of Particle Diameter and Magnetic Anisotropy on Magnetorelaxometry and Magnetic Particle Imaging Performance of Immobilized Magnetic Nanoparticles

Magnetic nanoparticles (MNPs) are of interest for application in magnetic particle imaging (MPI), an emerging biomedical imaging technology that relies on the non-linear dynamic magnetization response of MNPs. Compared to other molecular imaging modalities, MPI has the advantages of not requiring ionizing radiation and having high image contrast (zero signal from the tissue background), high sensitivity, and zero signal depth attenuation. In MPI, a static bias magnetic field gradient is superimposed with a uniform alternating magnetic field (AMF) to generate a small "field-free region", where the particles are able to respond to the AMF and generate a response signal. This signal is monitored by a pick-up coil and is then used to generate a quantitative image of the distribution of MNPs in a field of view. The variation of signal intensity with distance in the space around a point source in the imaging volume is commonly described using the so-called point spread function (PSF), which provides information on the signal strength and theoretical resolution obtainable with a given nanoparticle. A review of the recent literature suggests that no prior computational work has studied the effect of nature of magnetocrystalline anisotropy, strength of magnetocrystalline anisotropy energy, and particle diameter on the MPI performance (through the point spread function) of magnetic nanoparticles undergoing Néel-relaxation. On the other hand, prior computational studies have applied simulations of the Landau-Lifshitz-Gilbert (LLG) equation to predict the magnetorelaxometry of MNPs. However, these did not provide insight into the dipole dynamics for nanoparticles that are fixed in a physical matrix, such as would often be the case for nanoparticles that accumulate inside cells. Thus, further work is needed to fully understand how nanoparticle properties influence non-linear magnetization dynamics of nanoparticles, particularly in the context of the fields often used in MPI.

Here we report a computational study of the effect of particle diameter and magnetic anisotropy (considering both symmetry and energy) on the magnetization dynamics and MPI performance of spherical single-domain MNPs that relax by internal dipole rotation, by carrying out simulations that are based on the LLG equation. In the case of an applied static magnetic field, we studied the magnetic relaxation time of the nanoparticles as the field is suddenly applied and suppressed, for particles with uniaxial and cubic magnetic anisotropies and various anisotropy constants and particle diameters. The results, for both anisotropy symmetries show that for both cases where a static magnetic field is suddenly turned on or off, the MNPs may undergo a successive two-step (periods τ_1 and τ_2 for field turned on, periods τ_3 and τ_4 for field turned off) or combined one step (period τ_{12} for field turned on, period τ_{34} for field turned off) relaxation. Whether a particle relaxes with one or two periods when the field is turned on is independent of the particle size, but determined by the competition between the energy due to the applied magnetic field and the magnetic anisotropy energy. For the case where the applied magnetic field is suddenly turned off, our results show good agreement with theoretical predictions for the cases of $|K|V/k_B T \leq 1$ and $|K|V/k_B T >> 1$, where K represents the magnetocrystalline anisotropy constant, V represents the particle volume, k_B is the Boltzmann constant and T represents the absolute temperature. On the other hand, for the case of a dynamic applied AMF that is typical of MPI applications, the magnetic responses of the magnetite nanoparticles of various diameters were studied in terms of the magnetization dynamics, dynamic hysteresis loops, harmonic spectra, and PSF of the magnetization signal as a function of particle diameter. Results suggest that increasing the particle diameter leads to improved MPI performance (signal and resolution). Comparison of the MPI performance was also made between nanoparticles with uniaxial and cubic anisotropies. In summary, these computational studies provide theoretical insight into the role of particle diameter and magnetic anisotropy on the performance of MNPs for applications in MPI.

569135 DNA-Caged Polymer Nanocomposites for Erasable Fluorescence Imaging

Purpose: DNA follows well-known base pairing rules that allow for the creation of novel structures, including DNA origami. DNA origami can be designed into a wide variety of two- and three-dimensional shapes for a variety of applications, such as nanoscale engineering and drug delivery. Here, we report DNA cages, DNA nanostructures designed to form a network on surfaces. This DNA nanostructure was originally designed by Kurokawa et al.[1] for use as an artificial cytoskeleton inside liposomes. As a cytoskeleton, DNA cages increased the mechanical integrity of the lipids. Our goal is to integrate DNA cages with known micelle structures formed from block copolymers for applications in imaging, including erasable labeling of biological samples. Current procedures for multiplexed imaging via erasable labels require harsh treatment of tissue that can damage the samples. Multiplexed images allow for the labeling of multiple targets simultaneously with high image resolution, which means that complicated pathways in samples such as tumors can be studied using less time and materials. If the samples are damaged before all desired parts are labeled, additional samples may be required, which is not always possible when the samples come from a patient. Further, sequential sample labeling precludes evaluating the same markers in the same cells. The proposed labeling system is based on gentle DNA dehybridization interactions, which would greatly improve current technology.

Methods: The DNA origami previously described[1] was introduced to polymer micelles and solid polymer nanoparticles, forming a cage on their surfaces. Previously[1], it was suggested that DNA interacts with the surface of the micelles based on the charged group on the surface of the micelle and the negatively-charged DNA strands. Here, we show that attraction may also be based on the affinity of DNA for the polymer used. Proof of cage formation was measured based on the fluorescence of the cage itself, until a saturation point on the surface of the micelles was reached. DNA cages were tagged with single stranded DNA (ssDNA) targeting strands that reversibly attach to ssDNA strands labeled with fluorophores via strand displacement reactions. Strand displacement reactions are based on the highest number of complimentary base pairs. Thus, when an ssDNA with greater complement was added, the labeling strand was removed, erasing the signal.

Results and Implications: These data suggest that DNA cages have great promise for multiplexed cell labeling. Proper diagnosis of patients, especially when studying tumors, is highly dependent on cell labeling. DNA cages do not require harsh chemicals or procedures for removal, which may improve tissue labeling. The erasable nature of DNA cages allows for the labeling of the same target multiple times or different targets sequentially, potentially without damaging the sample. Finally, when combined with advanced image analysis techniques, it may be possible to create a three-dimensional image of the sample for a more complete analysis. Thus, DNA cages have the potential to significantly change cell labeling, especially that of patient samples.

Reference

1 Kurokawa, C. *et al.* DNA cytoskeleton for stabilizing artificial cells. *Proc Natl Acad Sci U S A* **114**, 7228-7233, doi:10.1073/pnas.1702208114 (2017).

572824 Tuning the Size and Composition of Manganese Oxide Nanoparticles for Magnetic Resonance Imaging of Breast Cancer

Magnetic resonance imaging (MRI) can detect more breast cancers than mammography, the current gold standard screening technique. However, MRI still produces a high false positive rate of up to 25% due to the commonly used gadolinium contrast agent, which is always active and nonspecifically highlights all vascularized tissue in the body. Poly(lactic-co-glycolic acid) (PLGA) nanoparticles containing manganese oxide (MnO) cores have been introduced as novel pH-sensitive contrast agents and can be activated upon tumor cell uptake into low pH endosomes to provide selective MRI contrast at the cancer site through Mn^{2+} generation. The objective of this study was to enhance MRI contrast by reducing MnO core size. Smaller MnO cores should allow for maximum packing volume into PLGA nanoparticles and faster controlled release of Mn^{2+} through a greater surface area to volume ratio. MnO cores were fabricated through thermal decomposition of manganese (II) acetylacetonate using two different temperature ramps (10°C/min and 20°C/min) and aged at 300°C for increasing time periods of 5, 15, and 30 minutes. This study is the first to show that temperature ramping rate and aging time impact both the resulting size and chemistry of MnO cores. Synthesized MnO cores were characterized by transmission electron microscopy (TEM), X-ray diffraction (XRD), and Mn^{2+} controlled release at pH 5 (to mimic endosomal pH inside cells), 6.5 (to mimic the tumor extracellular space), and 7.4 (to mimic the blood). A faster temperature ramp and shorter aging time at 300°C reduced MnO core size to 23.4 nm ± 11.7 nm as shown by TEM. XRD spectra revealed that pure MnO was obtained with longer overall reaction times; however, shorter overall reaction times produced a mixture of MnO and Mn_3O_4. Compared to MnO, Mn_3O_4 is expected to have slower Mn^{2+} production rates. Surprisingly, all synthesized cores had ~46% dissociation to Mn^{2+} over 24 hours in pH 5 media mimicking endosomes. Minimal Mn^{2+} release was found at pH 6.5 and 7.4, indicating that MnO nanoparticles must be targeted intracellularly to obtain sufficient MRI contrast for breast cancer detection. Future studies will investigate the impact of changing the ratio of chemical reactants to achieve further MnO size reduction and decreased polydispersity, as well as develop surface targeting methods to enhance MnO nanoparticle endocytosis within human breast cancer cells.

561685 Nanowires through Oriented Attachment or Nanoscale Self-Assembly

Metal nanowires with ultrathin diameters can be synthesized through oriented attachment or self-assembly without the involvement of any template. The success of such a synthesis relies on the use of a proper combination of precursor and reducing agent that can generate a large number of extremely small nanoparticles. Due to a quick depletion of precursor, the small nanoparticles are unable to grow in size through atomic addition. In the case of poor stabilization, the small nanoparticles automatically go through coalesce into ultrathin nanowires with a wavy morphology *via* oriented attachment. Thanks to the unique structure and involvement of planar defects, the as-obtainedultrathin nanowires typically show catalytic performance greater than the conventional catalyst based on supported nanoparticles This mechanism not only offers a powerful route to the synthesis of nanowires through attachment-based growth but also opens the door to the rational design and fabrication of novel metal nanostructures with enhanced properties.

559509 Directed Assembly of Paramagnetic Colloids in Time-Varying Magnetic Fields

Micron-sized superparamagnetic colloids placed in time-varying magnetic fields can provide us with rich dynamic behavior depending on the magnetic field applied and the particle concentration. In symmetric magnetic fields, a long-range attractive potential is induced resulting in the formation of colloidal clusters, which assemble into circular shapes that exhibit fluid-like behavior at relatively low fields, while becoming crystallized and highly ordered at higher field strengths. The self-assembly of these “colloidal atoms” has led to new insights regarding the dynamics phase transitions in two-dimensions. In particular, we have examined how tuning the “softness” of the interaction potential leads to interesting interfacial dynamics, as shown by coalescing clusters.These systems exhibit interesting coarsening and spinodal decomposition structural transitions for larger particle systems.

By utilizing a time-varying fields, interesting far-from-equilibrium behavior is observed. The questions addressed here involve identifying the effects of complex time-varying field conditions on the assembly structure and dynamics.

577027 A New Class of Colloidal Gel Based on the Solvent Segregation Assisted Particle Self-Assembly

In this work, a new type of thermally reversible gel is formed by a bottom-up approach using nanoparticles dispersed in a binary solvent similar to the well-studied bicontinuous interfacially jammed emulsion gel (bijel). However, unlike the particles in bigel that do not like either component of the binary solvent, nanoparticles in this new gel are prepared with a strong preference to one component of the binary solvent. As a result, preferential wetting to one liquid drives the particles into one liquid domain that is rich in one of the solvent components. The structure arrest of particles in one liquid domain cause the gel formation of the whole system, which is thus named as solvent segregation driven gel (SeedGel). The gelation mechanism is demonstrated by contrast matching experiments using small angle neutron scattering (SANS). The average domain size and gelation temperature have been also determined. It is found that the jammed nanoparticles render strong mechanical strength to the SeedGel. Free-standing nanoporous bulk materials that are centimeter in size could be obtained in different shapes by this method. The shape can be maintained even after removing the solvent from the SeedGel by heating. Dried SeedGel samples can be obtained without any micrometer or nanometer sized cracks as investigated by electron scanning microscopy (SEM). We believe that the SeedGel preparation method opens a door to self-assemble nanoparticles into porous materials, which is simple, yet useful for applications that require high surface areas.

575406 Adsorption Dynamics of Polymeric Nanoparticles at an Air-Water Interface with the Addition of Surfactants

The self-assembly of nanoparticles (NPs) at an air-water interface is greatly affected by the interaction between NPs and existing components on the interface. Previous studies have been conducted to model the adsorption of NPs onto an air-water interface. However, few studies in the past have quantitatively described the changes of NP adsorption dynamics due to the addition of reagents such as surfactants. In this work, NPs with hydrophobic cores and hydrophilic coronas are made using Flash NanoPrecipitation (FNP) method. We demonstrate that even at a very low concentration, ionic surfactants such as sodium dodecyl sulfate (SDS) and cetrimonium bromide (CTAB) can change the electrostatic potential of both the NPs and the interface, so as to affect the rate of adsorption of NPs at the interface. Quantitative comparison has been carried out to compare this electrostatic interaction energy with the change of adsorption energy derived from the Ward and Tordai model. We have also observed that the variation of NP adsorption due to the ionic surfactants can be eliminated by applying the shield effects of free salts, or by changing the core materials of NPs. Out findings have brought new insights into the fundamental understandings of NP interfacial assembly dynamics.

570967 Mechanisms of Nanoparticle Haloing for the Control of Colloidal Stability

Nanoparticle haloing is a mechanism by which binary colloidal suspensions can be stabilized as a result of particle size and charge asymmetry. As a model system, we have chosen negligibly charged silica microparticles mixed with low concentrations of highly charged zirconia nanoparticles. AFM measurements indicate an increase in the effective surface charge of silica particles with increasing zirconia nanoparticle concentration. This enhanced electrostatic repulsion between silica particles increases colloidal stability, which we attribute to nanoparticle haloing. We have characterized our model system over a range of conditions and discovered a narrow window of pH and nanoparticle concentration under which nanoparticle haloing is observed. Upon settling, colloidal packing was analyzed, and haloing was found to greatly affect equilibrium structures. DLVO model parameters were developed based on experimental measurements. Equilibrium properties of the binary system of silica and zirconia were evaluated by molecular dynamics simulations using HOOMD-blue. The free energy of microparticle association in various concentrations of nanoparticles was calculated from umbrella sampling simulations. Simulations predict the formation of loosely associated nanoparticle multilayers near the surface of each microparticle which we hypothesize leads to kinetic stabilization of the colloidal suspension in agreement with our experimental observations.

570589 Mechanistic Understanding of the Nanoparticle Self-Assembly By Coupling Inter-Particle Forces from Real-Time Measurements

Superlattice structures formed by nanoparticle (NP) self-assembly have attracted increasing attention due to their potential as a novel class of nanomaterials with enhanced physicochemical properties tailored by the assembly structure.[1] For instance, rationally designing superlattice patterns with defined separation distances can play an important role in controlling properties such as surface plasmon resonance. However, many key questions remain regarding the correlation between the dynamics of individual NPs and the emerging superlattice patterns. Such self-assemblies directly result from a subtle balance between various forces between NPs over different length scales. Brownian force (F_{Br}) resulting from thermal energy triggers motions of NPs. The van der Waals force (F_{vdW}) between NPs is generally a major attractive force at close separations, whereas the hydrodynamic force (F_D), electrostatic force (F_{elec}), hydration force (F_{hy}), and steric hindrance force (F_{Sh}) are repulsive forces.

To gain insights into the self-assembly process, we conducted in situ liquid cell TEM (LCTEM) experiments and monitored the silver or gold NP assembly process in real time to reveal the kinetics and the driving sources of assembly process under various experimental conditions, such as solvents and ligands.

To understand kinetic details of the process, the self-assembly of gold NPs are further analyzed by calculating the contributions of various forces involved [3-4]. Our results show that these forces, including F_{Br}, F_{vdW}, F_{el}, and F_D, and their interplay play a key role in NP self-assembly process. The superlattice structure can be controlled *via* solvents or ligands.

These results provide insights into the complex competition between forces at different scales. The obtained knowledge of interactions between surface coating materials can be extended to a wide range of ligands with different chain lengths, functionalized groups, or biodegradable polymers that can be applied to the fields of material synthesis, bio-mineralization, bio-sensors, or electronic devices.

References:

[1] M Boles, M Engel, and D. V. Talapin. Chemical reviews 116, (2016), p11220-11289.

[2] J. M. Yuk, J. Park, P.Ercius, K. Kim, D. J. Hellebusch, M. F. Crommie, J. Y. Lee, A. Zettl,; A. P. Alivisatos, , Science 336, (2012), p61-64.

[3] W.B. Russel, W B. Russel, D. A Saville and W. R. Schowalter, in "Colloidal dispersions", ed G.K. Batchelor (Cambridge university press, New York) p3-22

[4] J. N. Israelachvili, in "Intermolecular and surface forces", Academic press, (Elsevier, Santa Barbara) p3-22

573563 Probing Nanocrystal Shape Transformations Using Parallel Tempering Molecular Dynamics and the Mold Method

Solution-phase syntheses have produced an astounding variety of metal nanocrystal shapes with unique properties that depend on shape. Studies have shown that nanocrystal shapes often depend on the shape of the "seeds" – crystallites in the 1-10 nm size range that form subsequent to nucleation. The shapes of seeds (single crystal and twinned) are highly fluxional and sensitive to the solution environment surrounding them. Though many studies have probed minimum-energy shapes for small clusters in vacuum, very few studies have probed the effect of temperature and solution environment.

Focusing on Ag nanoclusters described by an embedded-atom method potential, we use parallel-tempering molecular dynamics simulations to generate ensembles of equilibrium shapes for Ag nanoclusters in the 2-10 nm size range as a function of temperature. We apply Common Neighbor Analysis to identify the crystallographic environments of cluster atoms and we create barcodes to characterize cluster structures. These studies show that clusters fall within a few structural motifs that can change with temperature. Interestingly, a solution environment lowers the cluster interfacial free energies and alters the preferred shapes.

To quantify free-energy differences between various structures, we employ the "mold method". In this method, we transform a nanocrystal from one shape to another by turning on and off the interactions between metal atoms and molds with initial/final shapes, while at the same time weakening/strengthening metal-atom interactions. This reversible transformation allows us to perform thermodynamic integration along the pathway, which yields the free-energy change. We create phase diagrams for clusters as a function of size, temperature, and solution environment. Our studies highlight the significant role of solution environment on nanocrystal shape.

582913 Interactions of Engineered Nanomaterials with the Cell Plasma Membrane

Most biomedical applications of nanotechnology require that engineered nanomaterials come into direct contact with mammalian cells. The cell plasma membrane, a lipid bilayer surrounding all cells, is the first cellular entity that interacts with incoming nanomaterials. Depending on dose, and nanoparticle and membrane properties, engineered nanomaterials are capable of disrupting the cell membrane, thereby causing cell toxicity. Due to the complex structure of the cell membrane, mechanistic research on nanoparticle-membrane interactions has primarily relied on membrane models. However, while a variety of nanoparticle properties have been examined, the structural complexities of the plasma membrane such as the compositional asymmetry of membrane leaflets and the presence of membrane domains have been overlooked. This has led to a lack of understanding on the mechanisms of nanoparticle-induced cell membrane damage. In this talk, I will discuss efforts in my group to develop membrane models that capture the complexities of the cell membrane and use these models to uncover the mechanisms by which nanomaterials disrupt membrane structure and integrity. These efforts demonstrate the important role of nanoparticle physicochemical properties and biomolecular corona, as well as membrane composition and biophysical properties in regulating nanoparticle-induced membrane damage. I will also provide an overview on our work on going beyond simple membrane models and utilizing live cells as the “model” to probe nanoparticle-membrane interactions.

582914 Shape-Selective Growth of Nanoscale Materials: Insights from Multi-Scale Theory and Simulation

A significant challenge in the development of functional nanomaterials is understanding the growth and transformations of colloidal metal nanocrystals. From a practical perspective, a knowledge of how to selectively synthesize desired metal nanocrystal sizes and shapes would benefit numerous applications. For example, nanocrystal synthesis science is playing an increasingly prominent role in electrocatalysis for fuel-cell applications, photocatalysis for the production of solar fuels, and the catalytic production of biofuels from biomass. Metal nanocrystals figure prominently in solar-cell technologies, for example as nanometallic plasmonic structures to enhance the light absorption and efficiency of photovoltaics, as non-plasmonic photosensitizers for light-energy conversion, and as nanowire-based transparent conducting films. Metal nanocrystals are also featured in emerging applications targeting energy efficiency, such as photothermic desalination based on plasmonic nanocrystals, wearable triboelectric generators, flexible energy storage devices, electrochromic smart windows, and life-inspired nanosystems. For these and many other applications, the science of shape-selective nanocrystal synthesis has been advancing at an increasingly rapid pace, with numerous recent reports on the synthesis of various beneficial nanocrystal morphologies.

Despite ample demonstrations that it can be highly beneficial to tune nanocrystal morphologies for specific applications and despite the tremendous strides made in nanocrystal synthesis science, it is still difficult to achieve high, selective yields in most synthesis protocols. Many fundamental aspects of these complex syntheses remain poorly understood and empiricism still runs rampant. Our research has focused on understanding shape-selective nanocrystal growth via multi-scale theory and simulations.

I will discuss our efforts to understand the thermodynamics and kinetics of shape evolution for Ag and Cu nanocrystal growth in solution. Our multi-scale theoretical studies based in quantum density-functional theory (DFT) highlight how various nanocrystal shapes can be governed by either thermodynamics or kinetics. We use a variety of classical molecular-dynamics (MD) simulation techniques based on our many-body force field to show that the growth of sufficiently large Ag nanocubes with PVP capping molecules is induced by the facet-selective deposition kinetics of solution-phase Ag atoms/ions. When chloride is introduced to the synthesis, Ag nanocubes grow with a thermodynamic driving force. Fivefold-twinned Ag nanowires grow by surface diffusion induced by the unique, strained structure of these fascinating objects. I will also discuss the growth of fivefold-twinned Cu nanowires in the presence of chloride and HDA, where DFT calculations indicate that growth is dominated by deposition kinetics. These scenarios indicate the various "knobs" that can be turned to achieve shape-selective syntheses in the future.

574588 Graduate Student Award: Investigating the Properties of Nanoparticles Containing Fullerene-like Molecules

Investigating the properties of nanoparticles containing fullerene-like molecules

Kimberly Bowal[1], Jacob W. Martin[1,2], Laura Pascazio[1], Markus Kraft[1,2,3]

[1] University of Cambridge, [2] Cambridge Centre for Advanced Research and Education in Singapore, [3] Nanyang Technological University

The structure and properties of nanoparticles containing curved polycyclic aromatic hydrocarbons (cPAHs) are investigated using molecular modelling. Due to the inclusion of one or more five-membered rings within their hexagonal lattice, these polar fullerene-like molecules possess unique steric and electronic properties that make them good candidates for many applications including microporous materials for gas storage, organic electronic devices such as imaging probes and batteries, sensors, and micelles for targeted nanomedicine. Development of these applications requires an understanding of the self-assembly and dynamic nanostructure of cPAH nanoparticles, which has not yet been well explored.

This work uses replica exchange molecular dynamics to determine the energy-dependent nanostructure of cPAH particles. The interactions between cPAH molecules are described using the new curPAHIP potential[1]. A range of cPAH sizes and ratios are investigated, along with systems containing planar PAHs and ions, to explore the homogeneous and heterogeneous particle internal arrangements and properties.

Homogeneous cPAH particles are found to be tightly packed and less stacked than their planar PAH counterparts. The constituent cPAH size plays a role in the molecular arrangement, with large 15 ring cPAHs ($C_{42}H_{14}$) displaying long-range order absent in nanoparticles containing small 7 ring cPAHs ($C_{20}H_{10}$). Heterogeneous particles containing cPAHs of different sizes, such as those shown in Figure 1, display a core-shell structure in which the larger molecules make up the core and the smaller molecules comprise the shell, as seen with planar PAH nanoparticles. This trend is not strongly influenced by the molecule ratio. In addition, the presence of planar PAHs and ions within cPAH nanoparticles promotes distinct arrangements dominated by weak dispersive interactions and strong electrostatic interactions, respectively. These systems may be indicative of nanoclusters generated in processes such as combustion.

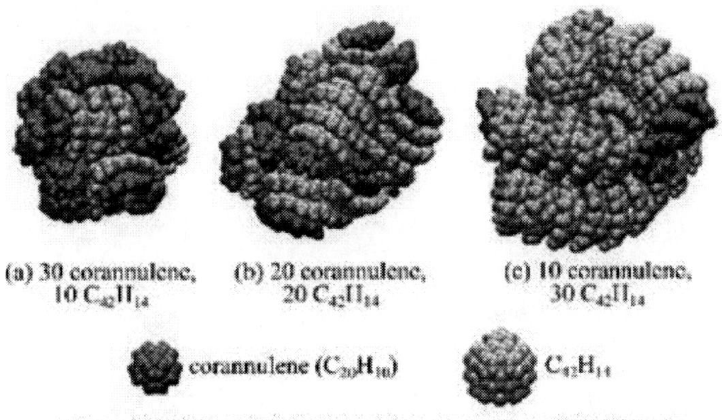

Figure 1: Stable configurations of heterogeneous cPAH clusters containing 40 molecules of corannulene and $C_{42}H_{14}$ molecules in different molecular ratios. Smaller corannulene molecules are coloured green and larger $C_{42}H_{14}$ molecules are coloured grey.

These results provide new information on structures and properties of nanoparticles containing cPAHs, including density, molecular structure, surface composition, and melting point estimation. This provides valuable understanding of the self-assembly and growth of nanoparticles with great potential in many applications.

[1]Bowal, K, Martin, J W, Misquitta, A J, Kraft, M: *Ion-induced soot nucleation using a new potential for curved aromatics,* 2019, Comb Sci Tech, doi: 10.1080/00102202.2019.1565496

574727 The Impact of Carbonaceous Nanoparticles to Climate through Their Composition and Morphology

Combustion-generated carbonaceous (soot) nanoparticles have the third strongest contribution to global warming and at the same time the highest uncertainty among all such contributors. Even though soot is made of agglomerates of polydisperse single and chemically-bonded (aggregated) primary particles, its optical properties are still estimated by most climate models using Mie theory for spheres neglecting soot morphology and composition and impeding the accurate estimation of soot direct radiative forcing and climate impact.

Here, the Rayleigh-Debye-Gans (RDG) theory is revised using discrete element modeling for surface growth and agglomeration [2] and discrete dipole approximation [3] to determine the direct radiative forcing of soot agglomerates accounting for the composition, necking (aggregation) and polydispersity of their constituent primary particles. For example, nascent soot single or aggregated primary particles identified in the nucleation mode of engine emissions (mobility diameter, d_m = 7-30 nm) have up to 35 % smaller absorption cross sections than large mature soot agglomerates found in the accumulation mode (d_m = 50-300 nm) due to their smaller C/H ratio. Furthermore, neglecting the fractal-like soot morphology results in 66 % smaller absorption and 6 times larger scattering cross-sections than those of mature soot agglomerates. The nascent and mature soot scattering/absorption cross-sections derived here by the revised RDG theory accounting for primary particle composition, aggregation and polydispersity are in excellent agreement with light scattering [4] and laser induced incandescence measurements in premixed flames and result in an average direct radiative forcing of 0.63 ± 0.05 W/m^2. This estimate is within those of employed in climate models but its uncertainty is 90 % lower than that from Mie theory. Employing this value, the soot contribution to global warming is reduced to 27.5 ± 2.5 % compared to the 19-58 % given today by climate models.

References:

[1] Kelesidis, G. A., Goudeli, E., & Pratsinis, S. E. (2017). Carbon 121, 527-535.

[2] Kelesidis, G. A., Goudeli, E., & Pratsinis, S. E. (2017). Proc. Combust. Inst. 36, 29-50.

[3] Kelesidis, G. A., & Pratsinis, S. E. (2019). Proc. Combust. Inst. 37, 1177-1184.

[4] Kelesidis, G. A., Kholghy, M. R., Zuercher, J., Robertz, J., Allemann, M., Duric, A., & Pratsinis, S. E. (2019). Powder Technol., doi.org/10.1016/j.powtec.2019.02.003.

566452 Multimodal Study of Biological Corona Structure and Dynamics on Single-Walled Carbon Nanotubes

Adsorption of polymers on single-walled carbon nanotubes (SWCNTs) has enabled developments in molecular sensing, *in vivo* imaging, gene and drug delivery, and chirality sorting applications. Noncovalent functionalization offers a modification route that preserves the pristine SWCNT atomic structure, thus retaining the intrinsic near-infrared SWCNT fluorescence for sensing or imaging functions, and offers a reversible binding mode for cargo delivery or chirality separation processes. However, noncovalently functionalized nanoparticles are inherently difficult to study due to the transient conformation of the solubilized complex. Moreover, noncovalent adsorption is a dynamic process, where exchange occurs between molecules in the bulk solution and molecules on the nanoparticle surface, into what is known as the 'corona phase'. Prior studies of the SWCNT corona[1,2] require drying or surface-immobilizing the complex and thus changing the native corona conformation, or do not operate at the necessary spatiotemporal scale to capture time-dependent binding behavior. Yet, the intended end-use of polymer-SWCNTs is in aqueous environments, thus the nature, strength, and kinetics of polymers adsorbing in the SWCNT corona are important contributors to the success of polymer-SWCNT based technologies. Understanding these parameters is especially important for applying functionalized SWCNTs in biological environments, where native proteins compete with the original polymer to occupy the SWCNT surface. Binding of proteins to the SWCNT not only disrupts the intended functionality of the nanoparticle, but further leads to unpredictable biocompatibility and biodistribution outcomes[3].

We present multimodal characterization of (i) the protein corona composition on polymer-SWCNTs in relevant biofluids, (ii) the structural conformation of the solubilized polymer-SWCNT complexes, and (iii) the kinetics of protein adsorption to the polymer-SWCNTs in solution. Single-stranded DNA-wrapped SWCNTs (ssDNA-SWCNTs) are chosen as the polymer-SWCNTs of interest due to their broad relevance in sensing[4], neuro-imaging[5], and chirality sorting[6], though this platform is extendable to other nanoparticles. We determine the protein corona composition by a selective adsorption assay with characterization by protein mass spectrometry in both human blood plasma and cerebrospinal fluid. We next probe the morphology of ssDNA-SWCNT complexes using biological small angle x-ray scattering (BioSAXS). Differences in corona morphology are observed depending on solution ionic strength, ssDNA length, and exposure to key corona proteins. To characterize the dynamic exchange in the SWCNT corona phase, we develop a multiplexed fluorescence assay that enables real-time tracking of biomolecule adsorption and desorption events. Studies are conducted with systematic variation of the molecular entities (ssDNA, protein) and solution conditions, whereby binding profiles inform a kinetic model of the SWCNT system under exposure to different biological conditions. The binding kinetics are compared against an orthogonal platform monitoring solvatochromic shifting of the near-infrared fluorescent SWCNT spectrum as a proxy for SWCNT corona perturbations. The work presented herein develops an understanding of the fundamental corona exchange mechanism, contextualizes the nature of the ligand exchange process, and provides insight into performance of these SWCNT-based systems in biologically relevant, protein-rich conditions such as human blood plasma and cerebrospinal fluid.

References

[1] Bisker, G. *et al.* Protein-targeted corona phase molecular recognition. *Nat. Commun.* **7**, 10241 (2016).

[2] Safaee, M. M., Gravely, M., Rocchio, C., Simmeth, M. & Roxbury, D. DNA Sequence Mediates Apparent Length Distribution in Single-Walled Carbon Nanotubes. *ACS Appl. Mater. Interfaces* **11**, 2225–2233 (2019).

[3] Nel, A. E. *et al.* Understanding biophysicochemical interactions at the nano–bio interface. *Nat. Mater.* **8**, 543–557 (2009).

[4] Kruss, S. *et al.* Neurotransmitter Detection Using Corona Phase Molecular Recognition on Fluorescent Single-Walled Carbon Nanotube Sensors. *J. Am. Chem. Soc.* **136**, 713–724 (2014).

[5] Beyene, A. G. *et al.* Imaging Striatal Dopamine Release Using a Non-Genetically Encoded Near-Infrared Fluorescent Catecholamine Nanosensor. *bioRxiv* 356543 (2018). doi:10.1101/356543

[6] Tu, X., Manohar, S., Jagota, A. & Zheng, M. DNA sequence motifs for structure-specific recognition and separation of carbon nanotubes. *Nature* **460**, 250–253 (2009).

567913 Process Safety Analysis for $Ti_3C_2T_x$ Mxene Synthesis and Processing

MXenes are two-dimensional metal carbides with promising applications in energy storage and sensors. Guidelines for safe, scalable MXene synthesis are important due to ongoing efforts to scale-up production of these novel nanomaterials. Hazard identification before scale-up will allow for inherently safer designs prior to actual implementation. Here, we investigate hazards associated with MXene production, including MAX phase synthesis from raw materials, etching of the MAX phase to MXene clay, exfoliation to MXene nanosheets, and post-processing of MXenes with $Ti_3C_2T_x$ as a model species. The major hazards in MXene synthesis are the potential for dust ignition, runaway reactions, and toxic chemical exposure. Because the synthesis of MXenes is a multi-step process, safety guidelines for each step are evaluated, including preventive and mitigating measures, best practices, and emergency procedures and responses. This includes handling of combustible powders, exothermic reactions, and harsh chemical etchants. This study is intended to facilitate safer MXene synthesis across various levels of scale-up, from large laboratory batches to commercial production.

574068 Measuring the Accessible Surface Area within the Nanoparticle Corona Using Molecular Probe Adsorption

The corona phase, the adsorbed layer of polymer, surfactant or stabilizer molecules around a nanoparticle, is typically utilized to disperse nanoparticles into a solution or solid phase. Unfortunately, few methods can directly probe the structure of this corona phase due to issues with dilution, molecular mobility, and total radiation cross section. In this work, we introduce a Molecular Probe Adsorption (MPA) method for measuring the accessible nanoparticle surface area using a titration of a quenchable fluorescent molecule. As an example, riboflavin is utilized to measure the surface area of Au nanoparticle standards, as well as corona phases on single walled carbon nanotubes and on graphenes. A material balance on the titration yields the surface coverage parameters, including the ratio of the surface area to dissociation constant of the fluorophore, q/K_D as well as K_D alone. Total 14 different nanoparticle corona phases, on AuNP, SWNT and graphene are chosen and studied with the new method. Uncertainty and precision are discussed, as well as the correlation of these parameters across different experiments. This rapid, quantitative technique should prove useful to elucidating the structure-property relationships of the nanoparticle corona.

568957 Role of Functional Ligands in Graphene Oxide/$MoSe_2$ Heterostructure(HS): Effects on Doping, Excitons and Transport

An accurate control on doping of transition metal dichalcogenide system opens up new horizons of pertinence towards device application. It is observed that doping profile of the $MoSe_2$ can be subtly modulated upon attachment with graphene oxide (GO). In the present study, we have demonstrated a precise control on the doping by manipulating interlayer coupling, using several first principles techniques. The key findings of our investigations are enlisted as follows. 1) By using DFT calculations incorporating spin-orbit coupling, it is shown that the doping of $MoSe_2$ can be tuned from n-type to p-type by varying the types and concentrations of the ligands attached to the GO-layer. Analysis of the band-structure of the heterostructure implies that the gross band behaviour of $MoSe_2$ remains unchanged in the HS. Attachment of GO only changes the effective mass of the carriers in the GO-bands appearing at the midgap and thus controls the transport properties. The optical properties of the HS are fully controlled by $MoSe_2$; 2) Investigation of the noncollinear behavior of magnetic property indicates that with the increment of ligand concentration, the almost isotropic magnetic moments of the heterostructure (HS) increases in all three directions; 3) Comparison of static optical property corresponding to the partial density of state reveals that the optical bandgap of pristine $MoSe_2$ is ~1.5 eV due to direct interband transition from valance to conduction band. Upon formation of HS with GO, many midgap states have been generated due to mutual charge transfer induced buckling of the GO and $MoSe_2$ layer; 4) Investigation of excitonic properties by using time dependent DFT indicates that the optical property of $MoSe_2$ undergoes exciton quenching upon attachment of GO. We have also experienced band-gap renormalization behaviour for low-energy exciton peaks; 5) The transport properties of the HS undergoes substantial changes depending upon the nature of placement of metal contacts. Analysis of the device potential of the HS as channel material with lateral and vertical metal (Au) contacts is done with self-consistent quantum transport calculations, which reveals that vertical contacts are more capable of retaining the obtained doping pattern than the lateral ones. This observation can be very useful for future optoelectronic and nanoelectronic device applications of this system as phototransistors or photodetectors.

569752 Repetitive Pool Boiling Runs: A Controlled Process to Form Reduced GO Surfaces from GO with Tunable Surface Chemistry and Morphology

Wettability and wickability of a surface are the key properties that govern a variety of industrial applications. Understanding the effect of aging on the durability of the surfaces and subsequent changes occurring in their molecular structure is vital to maintain functionality. Pool boiling, or two-phase heat transfer is a technique in which maintaining surface properties is critical to maintain the consistent heat dissipation from heater surface. Pool boiling involves a phase change of liquid by absorbing a large amount of latent heat which allows the dissipation of large amount of heat while maintaining low surface temperatures. In the recent years, several carbon nanomaterials-based coatings have been exploited for pool boiling applications due to their high thermal properties. However, the effect of aging on durability of these carbon nanomaterial-based coatings still remain an unexploited field.

Here, we have developed novel graphene oxide-copper (GO-Cu) porous coatings by a multi-step electrodeposition technique owing to higher thermal properties of GO and copper particles. We studied the effect of repetitive pool boiling on changes in molecular structure, surface chemistry, morphology, and wettability of the coatings. We observed the transformation of GO-Cu to reduced-GO-copper (rGO-Cu) coatings with repetitive boiling that led to improved boiling performance. This improvement is attributed to the transition of hydrophilic GO to hydrophobic rGO due to the repetitive pool boiling, and increased surface roughness as a result of hydrophobic surfaces. The durability, robustness and reliability of the GO-Cu coatings was found to be higher than pristine-copper and just-copper coated surfaces. A higher performance demonstrated by GO-Cu coatings indicate their increased sustainability and practicability in numerous engineering applications.

579901 Carbon Nanotube Photoluminescence Solvatochromism in Biomedicine: Spectroscopy, Imaging, and Modulation

Modulation of the carbon nanotube photoluminescence band position has implications for applications such as sensing and imaging. Carbon nanotube photoluminescence solvatochromism, the hypsochromic or bathochromic shifting of nanotube emission, can be induced via changes in the local dielectric environment. We developed methods to modulate nanotube solvatochromic behavior to improve sensor responses, and we developed tools to observe solvatochromism within live cells and, recently, live animals.

We found that nanotube solvatochromism extends to variations of local electrostatic charge, induced by adsorption/desorption of polyelectrolytes on the nanotube surface, such as proteins and nucleic acids. We also found that dielectric and charge-mediated solvatochromism can be compounded to exacerbate the response. Upon triggered desorption of a polyelectrolyte, such as upon desorption of RNA upon hybridization, for instance, amphiphilic molecules can adsorb to the recently-uncovered nanotube surface. The adsorption displaces water to result in hypsochromic shifting, which can be additive, as polyelectrolyte desorption will also cause a hypsochromic shift. We found that the amphiphile may be a surfactant but, recently, we found that proteins can result in a similar response, facilitating the detection of viruses.

We also developed tools to image changes in nanotube emission in live cells (near-infrared hyperspectral microscopy) and, recently, live animals (preclinical, near-infrared hyperspectral imaging). Carbon nanotube solvatochromism-based sensors, in combination with spectral imaging tools, allow measurements of biomarkers, drugs, and metabolites in live cells and organisms, facilitating improvements in disease detection, drug development, and biomedical research.

575978 Xeno Nucleic Acid Nanosensors for Enhanced Stability

The omnipresence of salts in biofluids creates a pervasive challenge in designing sensors suitable for *in vivo*applications. Fluctuations in ion concentrations have been shown to affect the sensitivity and selectivity of optical sensors based on single-walled carbon nanotubes wrapped with single-stranded DNA (ssDNA-SWCNTs). We herein observe fluorescence wavelength shifting for ssDNA-SWCNT-based optical sensors in the presence of divalent cations at concentrations above 3.5 mM. In contrast, no shifting was observed for concentrations up to 350 mM for sensors bioengineered with increased rigidity using xeno nucleic acids (XNAs). Transient fluorescence measurements reveal distinct optical transitions for ssDNA- and XNA-based wrappings during ion-induced conformation changes, with XNA-based sensors showing increased permanence in conformational and signal stability. This demonstration introduces synthetic biology as a complementary means for enhancing nanotube optoelectronic behaviour, unlocking previously unexplored possibilities for developing nano-bioengineered sensors with augmented capabilities.

570937 Sequence and Aggregation State Determine Stability, Endosomal Processing, and Long-Term Fate of DNA-Wrapped Carbon Nanotubes in Mammalian Cells

Single-walled carbon nanotubes (SWCNTs) have attracted much attention for applications in biomedical imaging and sensing due to their photostable, environmentally responsive near-infrared fluorescence. Single-stranded DNA was found to non-covalently disperse SWCNTs in water while preserving their optical properties and providing a biocompatible wrapping suitable for cellular interactions. The hybridization of DNA-SWCNTs has been studied extensively to determine the relationship between oligonucleotide sequence and SWCNT properties *in situ*, however intracellular environments introduce harsh conditions that can change the identity of the hybrid nanomaterial, altering its intrinsic optical properties. Aggregation via protein-nanotube interactions adds complexity to the biological system, while also causing concerns for cellular toxicity. Here, we show that the stability of DNA-SWCNTs after internalization by mammalian cells is dependent upon the oligonucleotide sequence that wraps the nanotube. The relationship between sequence length, fluorescence stability, and uptake are examined to provide insight on intracellular mechanisms that require considerations in nanotube sensor design. We show that aggregation not only alters the expected rate of uptake into the cells but further affects the endosomal processing pathways. We propose a model for the uptake and expulsion of SWCNTs that accounts for the aggregation state and DNA wrapping. These findings provide fundamental understanding of the interactions between nanotubes and live cells which can be applied towards development of carefully engineered carbon nanotube sensors.

570819 Real-Time Detection of H_2O_2 Signaling in Wild-Type Plants with Nanobionics Approach

As a consequence of their sessile nature, plants have developed complex signaling mechanisms to cope with the highly dynamic environment that they live in. Recent studies revealed that in response to external stimuli, plants can utilize various signaling pathways for cell-to-cell communication to engage different metabolic and genetic machineries involved in the plant defense system. Among the many signaling molecules uncovered, hydrogen peroxide (H_2O_2) represents one of the most versatile signaling molecule in plants in response to environmental stresses such as mechanical wounding. Herein, we develop a nanomaterial-based approach, termed plant nanobionics, using single-walled carbon nanotube (SWNT)-based nanosensors to monitor the wound-induced H_2O_2 signaling pathway in plants. This approach enables non-destructive, real-time detection of changes in endogenous H_2O_2 level in plants at a standoff distance. We demonstrate the versatility of this method to investigate the dynamics between H_2O_2 and electric signal in both model and non-model plants, and show that the nanosensor readout can be paired with portable and inexpensive electronic device. A mathematical model describing the self-propagating mechanism of the wave will be presented. Our work provides a promising tool to elucidate the complex plant signaling pathways across species to aid botany research and precision agricultural applications.

563265 Fluorescent Single-Walled Carbon Nanotubes for Protein Detection

Single-walled carbon nanotubes (SWCNTs) have unique optical and physical properties, and they benefit from the ease of surface functionalization and biocompatibility. Semiconducting SWCNTs fluoresce in the near-infrared (nIR) part of the spectrum, which overlaps with the tissue transparency window. The SWCNTs fluorescence is sensitive to the environment, and depending on the surface functionalization, subtle changes in the proximity of the nanotube can result in significant spectral modulations. Hence, SWCNTs can be utilized as optical sensors enabling real-time optical detection. I will present two recent discoveries of protein detection using SWCNTs functionalized with variants of poly(ethylene glycol). Using high-throughput screening against a panel of human blood proteins, we have discovers sensors for the proteins fibrinogen and insulin. The recognition also occurs in serum environment, showing that the SWCNTs sensors work in this complex environment despite the potential nonspecific adsorption. These results open new avenues for synthetic recognition of biological macromolecules, and hold great promise for medical and clinical applications.

Reference

1. Gili Bisker et. al., ACS Sensors, 3(2), 367-377 (2018).
2. Gili Bisker et. al., Nature Communications, 7 (2016).

563983 Directed Evolution of the Optoelectronic Properties of Synthetic Nanomaterials

Directed evolution is a powerful approach for tailoring protein properties toward new or enhanced functions. This technique allows one to alter a protein in a directed manner even in the absence of a known relationship between protein structure and function. Several nanocomplexes, such as optical sensors based on DNA-wrapped single-walled carbon nanotubes (DNA-SWCNTs), similarly lack a known structure-function relationship. For example, DNA-SWCNT complexes exhibit variations in their fluorescence that depend on the sequence of the DNA wrapping. However, little is known about the relationship between the DNA sequence and its effects on the fluorescence properties of SWCNTs. In this study, we use directed evolution as a guided approach to tuning the optoelectronic properties of DNA-SWCNT complexes through DNA mutation. Applying this technique over two rounds of DNA mutagenesis and screening, we evolved a DNA-SWCNT sensor with a 56% enhancement in fluorescence intensity. Furthermore, this sensor shows retained selectivity and sensitivity towards the analyte of interest. Though we demonstrate this approach to improve sensor brightness, directed evolution offers far-reaching possibilities in the field that can also be applied to improve additional sensor properties such as sensitivity and selectivity.

564858 The Impact of Covalent Functionalization on Single-Walled Carbon Nanotube Sensor Fluorescence and Function

Single-walled carbon nanotubes (SWNTs) have emerged as promising signal transduction elements for biological imaging and sensing due to their notable fluorescence in the near-infrared (NIR) region with minimal light absorbance by water or scattering by tissue.[1] Recently, exciting developments in SWNT surface chemistry have shown the retention and improvement of fluorescence intensity in SWNT after covalent surface functionalization.[2] These chemistries provide chemical handles for specific attachment of molecules of interest to the SWNT surface while keeping the intrinsic NIR fluorescence of the SWNT intact for bioimaging and sensing applications. However, despite their unperturbed fluorescence, it is unclear whether functionalized SWNTs retain their ability for molecular sensing, which necessitates a modulation in SWNT fluorescence provided by the molecular recognition of a surface-adsorbed polymer. We compare the photophysical properties of functionalized SWNTs to their pristine counterparts, and show SWNT optical properties remain available for sensing applications, but only for certain sensing strategies. Molecular recognition provided by surfactant and phospholipid SWNT surface coatings retain their ability to detect molecular analytes such as fibrinogen and insulin. Conversely, DNA oligonucleotide surface coatings under-perform for molecular recognition of dopamine compared to sensors constructed from pristine SWNT. Lastly, we explore the application of functionalized SWNT as dual-functional nanoparticles with both targeting and sensing capabilities. Our work establishes the potential advantages and drawbacks of covalent SWNTs functionalization, despite preservation of SWNT fluorescence, and implications for applications in molecular sensing.

(1) Bonis-O'Donnell, J. T. D.; Page, R. H.; Beyene, A. G.; Tindall, E. G.; McFarlane, I. R.; Landry, M. P. Dual Near-Infrared Two-Photon Microscopy for Deep-Tissue Dopamine Nanosensor Imaging. *Adv. Funct. Mater.* **2017**, *27* (39), 1–10.

(2) Setaro, A.; Adeli, M.; Glaeske, M.; Przyrembel, D.; Bisswanger, T.; Gordeev, G.; Maschietto, F.; Faghani, A.; Paulus, B.; Weinelt, M.; et al. Preserving π-Conjugation in Covalently Functionalized Carbon Nanotubes for Optoelectronic Applications. *Nat. Commun.* **2017**, *8*, 1–7.

566671 Protein Corona Formation on Single-Walled Carbon Nanotubes for Applications in Biological Environments

Engineered nanoparticles such as single-walled carbon nanotubes (SWCNTs) are increasingly used for biological sensing, imaging, and delivery applications owing to their distinctive optical and physical properties. SWCNT-based probes operate at spatiotemporal scales necessary to capture information on chemical signaling such as neurotransmission in the brain[1], and can deliver functional biological cargoes to tissues for genetic tranformations[2]. The critical – and often overlooked – challenge with these nanoscale tools is understanding the fundamental mechanisms of interaction between the nanoprobe and the system they are designed to query. When a nanoparticle enters a biological system, the surface becomes coated with proteins to form the 'protein corona'. Binding of proteins to the nanoparticle not only affects the structure and function of the native biomolecules, but has the added consequences of affecting nanoparticle biodistribution, masking the intended imaging or delivery function of the nanoparticle, and unpredictably re-defining the nanoparticle chemical identity[3].

To apply nanoscale tools in biological systems, it is crucial to understand the mechanisms and outcomes of protein corona formation. We present results on protein corona composition and binding dynamics for our SWCNT-based probes upon exposure to two biological fluids of interest: human blood plasma and cerebrospinal fluid. A selective adsorption assay with characterization by protein mass spectrometry reveals the unique composition of nanoparticle coronae, with identification of key corona proteins. Variation of incubation conditions provides insight on thermodynamic factors and dominant interactions driving protein corona formation. Next, the dynamic exchange of biomolecules in the corona phase is studied by a combination of near-infrared and visible fluorescence monitoring of the SWCNT and fluorophore-labeled proteins, respectively. Ultimately, a 'designer corona' strategy will be implemented to cloak the nanoprobes against untimely clearance pathways, to impart targeting abilities on a cellular level, and to encourage desired biodistribution outcomes on an organism level. Probing the corona composition, driving forces of formation, and timescales upon which the protein corona forms will inform figures of merit for the predictive design and synthesis of nanotechnology-based tools to be applied in protein-rich environments.

References

[1] Beyene, A. G. *et al.* Imaging Striatal Dopamine Release Using a Non-Genetically Encoded Near-Infrared Fluorescent Catecholamine Nanosensor. *bioRxiv* 356543 (2018). doi:10.1101/356543

[2] Demirer, G. S. *et al.* High aspect ratio nanomaterials enable delivery of functional genetic material without DNA integration in mature plants. *Nature Nanotechnology* 1 (2019). doi:10.1038/s41565-019-0382-5

[3] Nel, A. E. *et al.* Understanding biophysicochemical interactions at the nano–bio interface. *Nature Materials* **8**, 543–557 (2009).

573504 Emerging Organ Models and Organ Printing for Regenerative Medicine

Engineered materials that integrate advances in polymer chemistry, nanotechnology, and biological sciences have the potential to create powerful medical therapies. Our group aims to engineer tissue regenerative therapeutics using water-containing polymer networks called hydrogels that can regulate cell behavior. Specifically, we have developed photo-crosslinkable hybrid hydrogels that combine natural biomolecules with nanoparticles to regulate the chemical, biological, mechanical and electrical properties of gels. These functional scaffolds induce the differentiation of stem cells to desired cell types and direct the formation of vascularized heart or bone tissues. Since tissue function is highly dependent on architecture, we have also used microfabrication methods, such as microfluidics, photolithography, bioprinting, and molding, to regulate the architecture of these materials. We have employed these strategies to generate miniaturized tissues. To create tissue complexity, we have also developed directed assembly techniques to compile small tissue modules into larger constructs. It is anticipated that such approaches will lead to the development of next-generation regenerative therapeutics and biomedical devices.

571530 3D Layered Nano- Micro- Fibre Printing Techniques for Biological and Electronic Applications

Building two-dimensional (2D) and three-dimensional (3D) fibrous structures in the micro-nano scale will offer exciting prospects for numerous applications spanning from sensors, to energy storage and tissue engineering scaffolds. Existing fibre spinning techniques have been tailored for producing micro-to-nano scale fibers, but can be restrictive in the design of macroscopic fiber architectures in 3D space, and the choice of fibre materials. To address these technological gaps, we report two new fibre spinning techniques: first, 3D-LEP which combines low voltage electrospinning (LEP) and additive manufacturing to pattern suspended fiber layers in multiple tiers and designable orientations; and secondly, an efficient inflight fluidic fibre deposition (IFD) process to produce core-shell conducting fibres. Using these techniques, we demonstrate biological applications utilising fiber topography to guide the assembly of cellular aggregates in a 3D culture context; and optoelectronic applications showing unconventional transparent, suspended fibre-array electronics and the construction of layered 3D electrical architectures.

570135 Putting 3D Biofabrication to the Use of Tissue Model Fabrication

Microphysiological systems are microfluidic three-dimensional miniature human tissue and organ models that recapitulate the important biological and physiological parameters of their *in vivo* counterparts. They have recently emerged as a viable platform for personalized medicine and drug screening. These biomimetic microtissues are anticipated to supplement the conventional planar, static cell cultures, and to bridge the gaps between the current pre-clinical animal models and the human body. In addition, multiple microtissues may be channeled together through the microfluidics in a similar manner they arrange *in vivo*, providing the capacity to analyze interactions among these models. In this talk, I will specifically discuss our recent efforts on developing a series of bioprinting strategies including sacrificial bioprinting, microfluidic bioprinting, and multi-material bioprinting, along with various cytocompatible bioink formulations, for the fabrication of biomimetic 3D tissue models. These platform technologies, when combined with bioreactors and bioanalysis, will likely provide new opportunities in constructing functional tissue and disease models for drug development, therapeutics screening, and precision medicine.

570597 3D Printing Functional Voxels: Fate and Transport of Quantum Dots Under Capillary Flow

Nanotechnology is a technological branch that focuses on nanomaterials with dimensions smaller than 100 nm. Nanotechnology has and continues to impact many industries: petroleum, materials, cosmetics, and pharmaceutics, etc[1]. More than 1,300 consumer products based on nanotechnology have entered the marketplace[2]. Some common nanomaterials include: silica, fullerol, alumoxane, carbon nanotubes, metal oxide nanoparticles, nanocrystals, and quantum dots.

For three dimensional printing (3DP) companies, the ability to integrate these nanomaterials into additive manufacturing is a desirable outcome. By incorporating these nanomaterials into 3DP, it allows for an even broader platform of materials. Expanding the materials portfolio empowers the consumer, enabling those individuals the potential to develop and engineer novel materials to exploit new functional properties.

Understanding the transport and fate of these nano-sized materials is vital as these materials may eventually contaminate the subsurface environment and become a threat to the watertable[3]. Concern arise as evidence proposes that some nanomaterials can negatively impact human health and the environment[2]. A fundamental understanding of the fate of nanomaterials, specifically nanoparticles, is imperative for both applications as well as environmental and health impact.

This presentation proposes a novel non-invasive technique with high temporal and spatial resolution for imaging nanoparticle transport and retention in polymeric powders used for 3D printing. Engineering on the voxel-scale is a multi-physics process involving the infiltration, transport, adsorption, and retention of engineered nanoparticles (ENP's) suspended in a carrier solvent. Fundamentally, solute transport and retention in porous media involves macroscopic fluid dynamics, mesoscale advection-dispersion processes, and competing microscale physiochemical forces and interactions (e.g. DLVO forces) that occur at the interfaces between the solute particles and the porous media[3–5].

1. Zhang, T. *et al.* Investigation of Nanoparticle Adsorption During Transport in Porous Media. *SPE J.* **20**, 667–677 (2015).
2. May, R., Akbariyeh, S. & Li, Y. Pore-Scale Quantification of Nanoparticle Transport in Saturated Porous Media Using Laser Scanning Cytometry. *Environ. Sci. Technol.* **(in review**, 6–12 (2012).
3. Torkzaban, S., Bradford, S. A., Wan, J., Tokunaga, T. & Masoudih, A. Release of quantum dot nanoparticles in porous media: Role of cation exchange and aging time. *Environ. Sci. Technol.* **47**, 11528–11536 (2013).
4. Bradford, S. A. & Torkzaban, S. Colloid Transport and Retention in Unsaturated Porous Media: A Review of Interface-, Collector-, and Pore-Scale Processes and Models. *Vadose Zo. J.* **7**, 667 (2008).
5. Goldberg, E. *et al.* What Factors Determine the Retention Behavior of Engineered Nanomaterials in Saturated Porous Media? (2017). doi:10.1021/acs.est.6b05217

570866 Radio Frequency and Microwave Heating of Preceramic Polymers and Its Application in Silicon Carbide Processing

Here, we demonstrate microwave and noncontact Radio Frequency heating methods for mold free processing of silicon carbides (SiC) via preceramic polymer (polycarbosilane) route using carbon nanotube susceptors as fillers. This results in volumetric heating of polycarbosilanes and has applications in 3D printing of SiC and fiber processing.

We show that polycarbosilanes have minimum response to RF and addition of electromagnetic susceptors like MWCNT results in rapid heating and curing of these polymers. We perform dielectric characterizations on the polycarbosilane-MWCNT composites at room temperature. We were able to achieve 95 % extent of curing for polycarbosilane-MWCNT composite using RF in range of 76-78 MHz at 100 W. Microwave pyrolysis of polycarbosilane-MWCNT composites results in formation of cubic SiC structure. We used noncontact RF heating method for 3D printing ink composed of polycarbosilane and SiC nanoparticles with iterative deposition and curing to form a cross-linked structure. We also report for the first time change in dielectric constant and ac conductivity of the 3D printing ink as it heats and transforms while curing. We have studied rheological properties of the ink and its shear thinning behavior. We have also shown that RF/microwave heating results in molecular heating and avoids puddle formation while curing polycarbosilane fibers to SiC fibers.

577247 Complex Multi-Layered Microparticle Laden Fixed Hydrogel Wound Dressing Fabricated Via 3D Printing for the Treatment of Thermal Burns

While burns are a common injury in society today, the damage that burns – especially partial-thickness and full-thickness burns – can cause to bodily function and appearance cannot be underestimated. While skin grafting remains as the gold standard for the treatment of partial and full thickness burns, autologous skin grafting suffers from the prerequisite of needing sufficient healthy skin from the patient while the risk of rejection of donor skin by the patient's body is a concern in allogeneic skin grafting.

As either an alternative or a complement to skin grafting, wound dressings can be applied to provide a conducive environment that promotes healing and minimizes microbial infection. With the emergence of 3D printing technologies, the creation of complex 3D structures from a myriad of materials can be achieved inexpensively and efficiently. In this study, we propose the development of multi-layered fixed hydrogel wound dressings using 3D printing for the treatment of thermal burns. To further aid the wound healing capabilities of this wound dressing, microparticles encapsulating relevant drugs such as topical antibiotics and analgesics are incorporated into the wound dressing. 3D printing strategies used in the production of these hydrogel wound dressings along with their structure and function will be discussed.

Through this study, we hope to present novel strategies in which the strengths of 3D printing can be leveraged to fabricate complex hydrogel wound dressing structures with better efficacy in the healing of thermal burn wounds.

582009 Stooped Nanohairs with Randomized Configuration By Flood Electron Beam Under Ambient Pressure

Research Interests: Current fabrication process for high aspect ratio nanostructures has limitations; realization of randomly stooping feature, and lack of applicability of organic materials, although they possess attractive applications, Here, we present a new fabrication process of randomly bended polymeric nanohair structure by irradiating the polymeric vertical nanohair structure with flood electron beam under ambient pressure. Also, the bending degree was controlled by increasing irradiation time. This behavior can be explained by scattering of flood electron beam. Due to air molecules, electrons are severly scattered, and collided with the hairs from random direction. These electrons causes the oxygen-assisted radiolysis of PUA which result in outgas of carbon dioxides and water and new hydrogen bonds, hence mass loss and volum shrinkage occurs. These facts are verified by fourier transform infrared spectroscopy and monte carlo simulation of electron trajectories. This noble, and simple method has promissing applicability of fabrication of nanostructured surface with random feature for diverse applications, such as antireflection and dry adhesion.

Teaching Interests: N/A

557873 Particle Atomic Layer Deposition of Yttrium Oxide for Hydrolysis Protection and Sintering of Aluminum Nitride Particles

Aluminum nitride (AlN) has a high thermal conductivity and electrical resistivity, making it an attractive material for high power LED heat sinks. However, AlN particles hydrolyze in the presence of water, forming aluminum oxide (Al_2O_3) impurities which reduce the thermal conductivity of the material. Consequently, AlN powder must be processed in organic solvents which increases the cost of AlN heat sinks. Typically, AlN parts are densified by adding a small amount of secondary phase (ie, sintering aid) to enable liquid phase sintering and scavenge Al_2O_3 impurities at the grain boundaries to enhance thermal conductivity. Sintering aids are usually added by mechanical mixing, where primary ceramic particles (AlN) are mixed with a small quantity of secondary phase particles which can lead inhomogeneities in the final microstructure. In this work, particle atomic layer deposition (ALD) was used to conformally coat AlN particles with a thin film of yttrium oxide (Y_2O_3) to protect the AlN particle surface from hydrolysis and to enable liquid phase sintering. Transmission electron microscopy was used to characterize the as-deposited Y_2O_3 film. Investigation of the hydrolysis behavior of uncoated and coated AlN powders showed that the Y_2O_3 nanofilm effectively hindered hydrolysis of the AlN particles. The densification behavior of the powders was investigated by push-rod dilatometry and showed that the addition of Y_2O_3 by particle ALD increased the final relative density by ~30% relative to pure-AlN. Scanning electron microscopy and X-ray diffraction were used to determine the microstructural homogeneity and secondary phase formation of the dense AlN ceramics. The addition of Y_2O_3 by particle ALD homogeneously disperses the Y_2O_3 prior to densification and may benefit both the hydrolysis and densification behavior of AlN powders.

559364 Quaternized g-Carbon Nitride As a High Performance Electrode Material for Supercapacitors

Quaternized g-Carbon Nitride as a High Performance Electrode Material for Supercapacitors

For the first time, quaternized graphitic carbon nitride (Q-gCN) was used as base material for the fabrication of supercapacitor electrodes. More specifically, Q-gCN was synthesized via reductive alkylation of graphitic carbon nitride (g-CN) and subsequently deposited onto a carbon fiber paper by paint brush method. A wide variety of characterization techniques including XRD, FTIR, Raman, UV-vis, TGA, SEM, and TEM were used to assess the physicochemical properties of these novel materials. In addition, the effect of some variables such as material load and electrolyte nature on the electrode performance were evaluated using cyclic voltammetry, charge-discharge test, and electrochemical impedance spectroscopy. For this purpose, a split test cell (MTI Corp.) containing two identical electrodes separated by a piece of cellulose paper was used as a supercapacitor. Before running the tests, some drops of H_2SO_4 (or KOH) electrolyte solution were added to the cell.

560440 Effects of Zeolite Dopant Atom Concentration on Noble Metal Nanoparticle Sintering Observed with *in Situ* UV-Vis Spectroscopy

Elucidating the mechanisms of catalyst deactivation is an important step in improving industrial catalyst lifetime, which will in turn lead to substantial cost savings via reduced operational downtime and reduced purchasing frequency of precious-metal-impregnated catalysts. Deactivation can be caused via sintering of catalytically active surface-supported metal particles. For this reason, we have investigated the behavior of zeolite-supported gold nanoparticles (ZEO@AuNPs) at high temperatures to better understand the role that zeolite dopant atom concentration and dispersion play in processes that lead to AuNP sintering. AuNP plasmon resonance position and intensity are sensitive to AuNP size and proximity. In leveraging this resonance, *in situ* UV-vis spectroscopy (corroborated by *ex situ* TEM) was used to track time-dependent changes in ZEO@AuNP size and abundance. These experiments indicated the operation of competing sintering mechanism components, namely Ostwalt ripening and particle coalescence. Our findings are complimented by computational models of AuNPs on zeolite surfaces, which suggest correlation between the ease of mobility for ZEO@AuNPs and the energetic landscape of the zeolite's surface. Data and analyses from experiments on these materials will be presented with context provided by computational models.

560590 Ni^{2+}–Modified g-Carbon Nitride Deposited on Carbon Fiber Paper for Supercapacitor Applications

In the present work, Ni^{2+}–modified g-carbon nitride (Ni-gCN) was used as base material for the fabrication of supercapacitor electrodes. More specifically, Ni-gCN was synthesized via cationic complexation and subsequently deposited onto a carbon fiber paper by paint brush method. Some characterization techniques such as XRD, FTIR, Raman, UV-vis, TGA, SEM, and TEM were used to assess the physicochemical properties of this novel material. In addition, the effect of some variables such as material load and electrolyte nature on the electrode performance were evaluated using cyclic voltammetry, charge-discharge test, and electrochemical impedance spectroscopy. For this purpose, a split test cell (MTI Corp.) was used as a supercapacitor. In this case, two identical electrodes (anode and cathode) where conveniently placed in the cell and separated by a piece of filter paper. Some drops of H_2SO_4 (or KOH) electrolyte solution were added to the cell before the tests.

565017 Engineering Xeno Nucleic Acid Nanosensors for Enhanced Stability

The omnipresence of salts in biofluids creates a pervasive challenge in designing sensors suitable for *in vivo* applications. Fluctuations in ion concentrations have been shown to affect the sensitivity and selectivity of optical sensors based on single-walled carbon nanotubes wrapped with single-stranded DNA (ssDNA-SWCNTs). We herein observe fluorescence wavelength shifting for ssDNA-SWCNT-based optical sensors in the presence of divalent cations at concentrations above 3.5 mM. In contrast, no shifting was observed for concentrations up to 350 mM for sensors bioengineered with increased rigidity using xeno nucleic acids (XNAs). Transient fluorescence measurements reveal distinct optical transitions for ssDNA- and XNA-based wrappings during ion-induced conformation changes, with XNA-based sensors showing increased permanence in conformational and signal stability. This demonstration introduces synthetic biology as a complementary means for enhancing nanotube optoelectronic behaviour, unlocking previously unexplored possibilities for developing nano-bioengineered sensors with augmented capabilities.

565654 Thermo-Responsive Ionophore Polymer Brushes Grafted on Magnetic Graphene Oxide As a Dual-Functional Composite Adsorbent for Selective Lithium Ion Recovery from Seawater

Lithium (Li) is one of the most important components for energy storage due to its high energy density. Sufficient amount of lithium is currently produced from ores and brines, but a supply shortage is anticipated due to its soaring demand in mobile electronics and electric vehicle industries. Furthermore, the uneven global distribution of lithium reserves necessitates its recovery from alternative lithium sources. Seawater is identified as a potential alternative lithium resource as it contains $2x10^6$ Mt Li^+. However, the complex composition of seawater makes it difficult to recover Li^+ due to its significantly lower concentration than Na^+, K^+, Mg^{2+}, and Ca^{2+}.

In this work, a multi-functional thermo-responsive, Li^+-selective adsorbent with magnetic property was successfully synthesized. The adsorbent used magnetic graphene oxide (rGO-Fe_3O_4) as a support, on which the polymer brushes of co-polymerized Crown Ether (CE) and N-Isopropylacrylamide (NIPAAm) were grafted via surface-initiated atom transfer radical polymerization (SI-ATRP). Magnetite (Fe_3O_4) is used for material separation and recyclability, and graphene oxide (GO) as a two-dimensional, high-aspect ratio support material for the polymer brushes and magnetite. The CE component is responsible for the selective Li^+ capture whereas the NIPAAm imparts the thermo-responsive character.

The GO support decorated with magnetite (rGO-Fe_3O_4) was synthesized through a modified Hummers' method and solvothermal process. This was followed by grafting hydroxyl moieties (OH-rGO-Fe_3O_4), via diazonium surface modification technique and further functionalized to immobilize the SI-ATRP initiator. On the other hand, Li^+-selective CE (i.e. 2-hydroxymethyl-12-Crown-4 Ether) was modified to introduce an allyl-functional group. The allyl-CE was mixed with NIPAAm and reacted with the ATRP initiator to produce the polymer brushes.

The synthesized composite adsorbent material (12CE4/NIPAAm@HCTP-rGO-Fe_3O_4) was characterized using Boehm Titration, Transmission Electron Microscopy (TEM), X-ray Diffraction Spectrometry (XRD), Thermogravimetric Analysis (TGA), Fourier-Transform Infrared Spectroscopy (FT-IR), Raman Spectroscopy, and X-ray Photoelectron Spectroscopy (XPS). UV-Vis was utilized to determine the Lower Critical Solution Temperature (LCST) of the polymer brushes which is critical for the control of the adsorption/desorption process of Li^+. The adsorbent material was systematically tested to determine the Li^+ adsorption capacity and adsorption kinetics. The selectivity of 12CE4/NIPAAm@HCTP-rGO-Fe_3O_4 was tested using seawater as the feed solution. Repeated material adsorption-desorption experiments were done by utilizing the thermo-response of the material for the capture and release of the captured Li^+ without the utilization of acid. Thus, the integration of both magnetite and the polymer brush on the GO support material resulted to the Li^+-selective, magnetic and thermal-responsive 12CE4/NIPAAm@HCTP-rGO-Fe_3O_4 which is suitable for long-term Li^+ adsorption application.

This research was supported by Basic Science Research Program through the National Research Foundation of Korea (NRF) funded by the Ministry of Science and ICT (No. 2017R1A2B2002109) and by the Ministry of Education (No. 2018R1D1A1B07048007 and No. 22A20130012051(BK21Plus)).

566348 DNA Reaction-Diffusion Attractor Patterns

Living systems possess the ability to form and recover complex patterns in prescribed locations at length scales of hundreds of microns. In this study, we use oligonucleotide reaction networks, and leverage digital maskless photolithography and microfluidic delivery methods to program the shape of biomolecular attractor patterns in photopatterned poly(ethylene-glycol) diacrylate hydrogels. Linear and 'hill'-shaped DNA-based patterns formed and stabilized in 4 and 9 hours respectively. When patterns were perturbed in particular locations with varying dosages of UV light, they reformed their equilibrium profile over times ranging from 1 to 10 hours.

567498 Near-Infrared Confocal Imaging of Single-Walled Carbon Nanotube Uptake in Bacterial Cells

The distinctive properties of single-walled carbon nanotubes (SWCNTs) have inspired the development of innovative applications in the field of cell nanobiotechnology. Most studies to date have focused on eukaryotic cells capable of internalizing SWCNTs functionalized with a variety of non-covalent wrappings. However, the effect of SWCNT functionalization on transport across the thick cell wall of prokaryotes remains unexplored. In this study, we explore the uptake of SWCNTs in Gram-negative cyanobacteria and demonstrate selective internalization of SWCNTs decorated with charged protein wrappings. The functionalized SWCNTs are shown to traverse the outer cell wall of both filamentous and unicellular strains of cyanobacteria, independent of whether the strain is naturally competent for DNA uptake. A custom-built, spinning disc confocal microscope was used for the first time to directly image near-infrared (NIR) SWCNT fluorescence within cells, revealing a highly inhomogeneous distribution of SWCNTs that is otherwise overlooked using conventional NIR widefield imaging. The nanobionic cells show sustained photosynthetic activity and growth, offering a powerful avenue for engineering photosynthetic organisms with augmented and even inherited nanobionic capabilities.

568045 Characterization of Double-Stranded DNA on Single-Walled Carbon Nanotubes (SWCNTs)

DNA has been extensively studied due to its versatility as a dispersant of SWCNTs. The specificity and binding affinity of DNA to various analytes has been shown to be strongly sequence dependent. As a result, through modifications of the nucleotide sequence, the DNA conformation on the SWCNT surface can be tailored to suit specific needs for techniques such as single-molecule detection, *in vivo* imaging, and chirality separation. To date, the majority of research has focused on the interaction between single-stranded DNA and SWCNT surface, with less focus on the nature of the interaction between double-stranded DNA (dsDNA) and SWCNTs. As a result, the exact interaction mechanism governing this type of complex remains strongly debated and largely unknown.

In this study, we employ various biochemical methods to infer the conformation of dsDNA on the surface of SWCNTs. Our methods are based on imaging techniques for identifying DNA-modifying enzyme activity in the presence of dsDNA-SWCNT complexes. Our findings suggest that dsDNA can partially retain its native conformation on the SWCNT surface, and the degree of dsDNA accessibility is strongly sequence dependent. These findings offer new possibilities for SWCNT sensing applications that employ dsDNA, such as the optical detection of DNA-protein interactions.

569258 Impacts of a Low-Temperature Calcination on the Characteristics of Monodisperse, Surface-Clean, Supported Nanoparticle Catalysts

Traditional methods of preparing supported nanoparticle catalysts require the use of stabilizing ligands that passivate the nanoparticle surface to reduce undesired morphology changes. While effective in maintaining monodispersity during synthesis, stabilizing ligands often significantly hinder catalytic activity as they compete with reagents for active sites and are therefore typically removed through a high-temperature annealing step prior to use. It is well documented that exposure to high temperatures results in significant nanoparticle growth ultimately resulting in polydisperse populations and reduced surface area for catalysis; however, the detrimental effects of thermal treatments on catalyst surfaces beyond those associated with size changes are typically overlooked. In this work, we demonstrate that even a low-temperature calcination has detrimental effects on catalyst properties including changes in nanoparticle morphology that result in a significant decrease in activity, a change in surface hydrophilicity, a change in activation energy, and results in the formation of an induction time when utilized in a reaction. While calcination remains a widely used method catalyst activation, the negative effects of high temperatures on catalyst properties are often overlooked.

571185 MOFs Nanosheet Formation in Microreactor

MOF nanosheets show better performance than nanobulk in the mixed matrix membrane for gas separation due to their ease of dispersity and small dimension. However, as reported in the literature, MOF nanosheet preparation methods are few in number, difficult to execute, and of low yield. Microreactors can realize diffusion of two flows without being affected by external interference. In this work, interface will be created by a T-union microreactor in order to imitate the situation in the 'interfacial method'. Particle size could be controlled by precise control of the mixing. Higher yield was realized because of the efficient diffusion provided by micromixing. The prepared nanosheets were used for catalysis of gas-liquid-solid reactions, and the results of using nanosheet or bulk were compared.

571248 Defect State Modulation of Wide Band Gap Semiconductor Nanoparticle for Potential Manifold Application

Surface modification of inorganic nanoparticle using different ligand is well-known to enhance different functionality of the nanoparticles (NP), which improves tuning of relevant properties for suitable applications in the field of nanotechnology. As the NPs have very high surface to volume ratio compare to its bulk counterpart, it has higher possibility of having surface defect. Surface defects of wide band gap semiconductors play a crucial role in their properties. Defect mediated recombination of photoinduced electron-hole pair in the semiconductors is very much well known. By modulating the defect state, it is possible to control the defect mediated recombination. Herein, we present a facile synthesis of ZnO nanoparticle followed by surface modification using halide ions by simple precipitation method and demonstrate that proximity of various halide ions to the ZnO NP changes their electronic properties. This kind of surface modification shows very little impact on its surface morphology. On the other hand, the functionalization of halide in the surface of ZnO plays a significant role to modulate its electronic property. This change of electronic property influences their applications in photocatalytic activity or magnetism. The halide attached ZnO NP displays a large reduction of defect state emission. In the present work, we have investigated that for wide bandgap semiconducting oxides, use of anionic halide attachment like Cl^- as surface defect healer proves to be more useful for photocatalytic application than bulk doping using cationic dopant like Mn. We have utilized various microscopic and spectroscopic tools (including picosecond resolved technique) to characterize the nanomaterial and the impact of different halide ions to regulate the light induced charge separation in ZnO NP. Picosecond resolved fluorescence transients reveal a significant annihilation of fluorescence decay, implying an electronic interaction between the NP and the attached halide ions. Upon attachment with a suitable halide ion, room temperature ferromagnetism can be achieved in the nanohybrid system. Finally, the photocatalytic activity of the halide-ZnO NP was carried out under UV light irradiation. To illustrate the impact of defect state and halide ion on the ZnO NP surface, we have carried out first principles DFT calculations using Vienna Ab-initio Simulation package (VASP). It is found that ZnO nanoclusters with surface oxygen vacancy produce trap states within the band gap of the nanoclusters. These states effectively confine the photoinduced electrons and thus essentially reduce the photocatalytic yield with respect to pristine ZnO. However, upon proper halide attachment to the defect states, it is possible to reduce the trap state of the NP recovering the efficacy of reactive oxygen species (ROS) generation in the aqueous solution. We have shown both experimentally and theoretically, that attachment of certain halide ions onto ZnO NP affects its electronic structure and gives rise to a ferromagnetic behaviour even in the absence of magnetic ions. So, the particular halide-ZnO (namely Cl-ZnO) consolidate higher magnetic recovery, thus can be easily separated from water after photocatalysis. In conclusion, modulation of surface defects using chloride can improve various properties of ZnO, demonstrating its utility for functionalization of ZnO NP.

571311 Fast and Green Microwave-Assisted Synthesis of Titania Nanotubes

Microwave-assisted synthesis possesses exclusive benefits of uniform, rapid, and volumetric heating and is able to significantly reduce reaction duration. Many researches have been done in microwave crystalline of titania nanotubes, however, most of them are batch synthesis which are limited to small-scale batch preparation. In this study, a fast and green microwave-assisted flow synthesis method for continuous production of titania nanotubes is developed. The synthesized materials were characterized by transmission electron microscopy (TEM), X-ray diffraction (XRD) and micro-Raman. The effects of different parameters, including microwave irradiation temperature, pressure, and residence time on the physicochemical properties of titania nanotubes were investigated.

571852 Bulk Synthesis of Quasi-Carbon Dot Via Two Step-Method from Rice Husk

Bulk synthesis of quasi-carbon dot via two step-method from Rice Husk

Prashant Kumar, Sudipto Chakraborty

Department of Chemical Engineering, IIT Kharagpur
E-mail: prashantkumar.ch@iitkgp.ac.in, sc@che.iitkgp.ernet.in

Abstract

Carbon dot (CD) – A newbie from the carbon nanomaterial family is gaining utmost attention, due to attractive property such as non-toxicity, solubility, stability[1]. Additionally, the oxygenous functional group on the surface imparts them reducibility property thereby ease of functionalization[2], so it can reduce metal precursors and forms compound with metals possessing high thermal conductivity[3]. Due to these properties, it is getting due relevance in heat transfer application, in particular, nanohybrid kind of nanofluid synthesis[3]. Therefore, the need for bulk CD for nanofluid synthesis arises. Herein, we propose a method of carbon dot synthesis using the two-step method for the bulk CD requirement, a method derived from CD synthesis from rice husk that finds application in other areas [4]. The proposed method involves carbonization followed by nitric acid oxidation, followed by a copious amount of distilled water wash leads to the formation of purified quasi-carbon dot from rice husk (CD-RH/CD). The synthesized CD was characterized using various characterization techniques. FTIR analysis was done to check the formation of associated functional groups of the CD. In figure 1, FTIR plot shows that oxidation of carbonized rice husk (C-RH) with nitric acid causes the introduction of carboxyl functional group at 1720 cm^{-1}, as well as enhancement in intensity of functional groups such as hydroxyl, aldehyde at their respective peak positions in CD, which confirms the formation of various oxygen-containing the functional group. CHNS analysis was done to check the changes in the elemental composition due to CD formation, which is compared with carbonized rice husk as shown in table 1. The increase in oxygen content in case of CD indicates the formation of the oxygen-containing functional group in species. Zeta potential was done to check to dispersion stability of C-RH and CD-RH in water and is tabulated in table 2. Increase in the zeta potential value can be attributed to the increase in oxygen functional group on carbon surface which imparts higher stability than carbonized rice husk. DLS was performed for size determination of the prepared nanoparticles. XRD was done to check the presence of the amorphous and graphitic structure in the sample. FESEM was done to study their morphology. Finally, EDAX was done to check the surface composition of different elemental species of the sample. Due to brevity, only few characterization results are included with this abstract. Above mentioned analysis is in concurrence with the one available in the literature for CD. The synthesized CDs have potential application in metal-based nanohybrid preparation, which can be used in heat transfer enhancement where water is present as the base fluid.

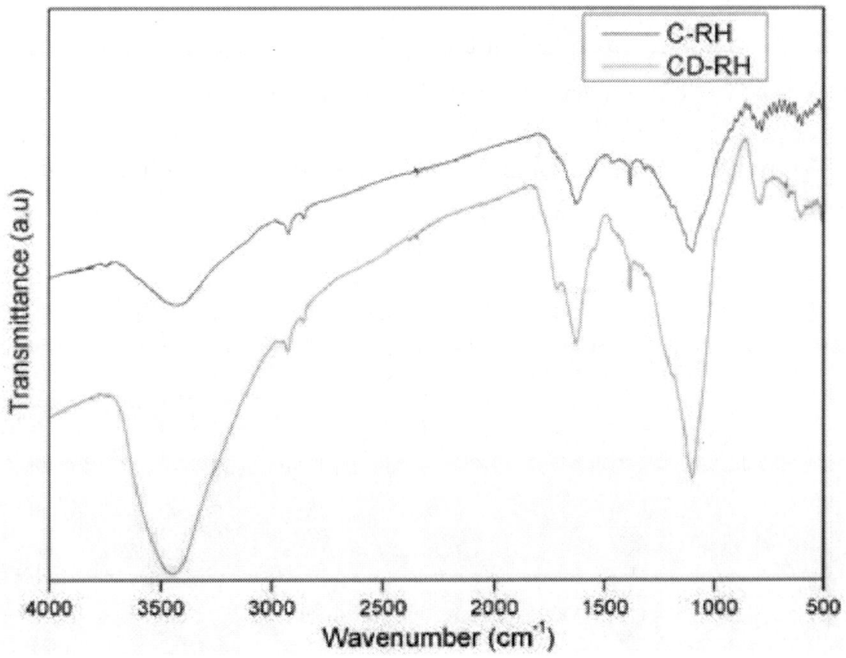

Figure 1. FTIR plot of carbonized rice husk (C-RH) and carbon dot derived from rice husk (CD-RH)

Table 1. Elemental composition of carbonized rice husk (C-RH) and carbon dot from rice husk (CD-RH)

Species	Element	C	H	N	S	O
C-RH		46.39	2.285	0.43	0.057	50.8
CD-RH		21.59	1.748	1.13	0.070	75.5

Table 2. Zeta potential of carbonized rice husk (C-RH) and carbon dot from rice husk (CD-RH)

Species	Zeta potential
C-RH	-38.6
CD-RH	-45.8

References:

[1] A. L. Himaja, P. S. Karthik, and S. P. Singh, "Carbon Dots: The Newest Member of the Carbon Nanomaterials Family," *Chem. Rec.*, vol. 15, no. 3, pp. 595–615, 2015.

[2] R. Das, R. Bandyopadhyay, and P. Pramanik, "Carbon quantum dots from natural resource: A review," *Mater. Today Chem.*, vol. 8, pp. 96–109, 2018.

[3] Z. Azizi, A. Alamdari, and M. Mohammad, Doroodmand, "Highly stable copper/carbon dot nanofluid Preparation and characterization," *J. Therm. Anal. Calorim.*, 2018.

[4] P. Zhen *et al.*, "Synthesis of carbon nanoparticles from waste rice husk used for the optical sensing of metal ions," *New Carbon Mater.*, vol. 31, no. 2, pp. 135–143, 2016.

572411 Polyacrylonitrile and Polypropylene Carbonate Blended Solid Polymer Electrolytes for Application Toward Solid State Lithium Ion Batteries

Lithium ion batteries are promising as energy storage devices because of their high energy density; however, current technology for lithium ion batteries is not safe for large scale applications and general use. Traditional lithium ion batteries use an organic liquid electrolyte to transport lithium ions between electrodes. Organic solvents used in lithium ion batteries, such as diethyl carbonate, are highly flammable and volatile. Safety issues arise when the battery heats from charging, discharging, or from dendrite formation in the cell. When this occurs, the battery can potentially ignite and pose a safety hazard. In this work, the research focus is developing blended solid polymer electrolytes for solid state lithium ion batteries. Polypropylene carbonate (PPC), an amorphous polymer, and polyacrylonitrile (PAN), a semi-crystalline polymer, have been shown to be individually effective for solid state lithium ion batteries. In this work, we fabricate novel PAN and PPC blended solid polymer electrolytes complexed with lithium bis(trifluoromethanesulfonyl)imide (LiTFSI). Furthermore, LiTFSI acts as both a source of lithium ions and as a plasticizer for the polymers, thus increasing the amorphous character of the polymer blend. Lithium ion conductivity is higher in amorphous electrolytes than in semi-crystalline solid polymer electrolytes. Solution casting, a commonly used technique for electrolyte fabrication, is used to control the thickness of the electrolyte film by varying the amount of solution added to the casting plate. During solution casting, the PPC and PAN undergo phase separation, this increases the number of micropores in the electrolyte. It has been shown that micropores improve the ionic conductivity of solid electrolytes. Upon evaporative drying of the solid polymer electrolytes, several tests are performed to characterize the electrolyte material as test well as battery performance.

573148 Real-Time Evaluation of Cellular Behavior upon Exposure to Metallic Organic Frameworks

Metallic organic frameworks (MOFs) are a type of nanoparticle made up of metal ions linked through an organic ligand. MOFs beneficial features such as a high surface area, mobility, and porosity, as well as, thermal and chemical stability make them desirable candidates for a vast array of applications from drug delivery to tissues engineering, and from biocatalysis to membrane separation systems. However, MOF implementation in such sectors could lead to toxic effects resulted from exposure of individual user during either manufacturing or particle usage.

Herein we proposed to advance our understanding of MOFs toxicity using both real-time and single point cellular assays. For this, we considered lung epithelial cells and ZIF-8, a lab-synthesized MOF as model systems and electric cell impedance sensing and colorimetric assays that assessed enzymatic functionality before, during and after cellular exposure. The above *in vitro* analysis complemented by microscopical observations helped evaluate viability and metabolic activity of the exposed cells and showed that there is a dose dependent toxic effect with identified time frames of action and subsequent cellular changes in adhesion or behavior. To our knowledge, this is a first combinatorial strategy used for MOF's toxicity determination. The derived understanding the MOFs' mechanisms of toxicity is envisioned to aid in minimizing particle deleterious effects to potentially lead to safe-by-design strategies when such material development is considered.

573706 Sequence and Aggregation State Determine Stability, Endosomal Processing and Long-Term Fate of DNA-Wrapped Carbon Nanotubes in Mammalian Cells

Single-walled carbon nanotubes (SWCNTs) have attracted much attention for applications in biomedical imaging and sensing due to their photostable, environmentally responsive near-infrared fluorescence. Single-stranded DNA was found to non-covalently disperse SWCNTs in water while preserving their optical properties and providing a biocompatible wrapping suitable for cellular interactions. The hybridization of DNA-SWCNTs has been studied extensively to determine the relationship between oligonucleotide sequence and SWCNT properties *in situ*, however intracellular environments introduce harsh conditions that can change the identity of the hybrid nanomaterial, altering its intrinsic optical properties. Aggregation via protein-nanotube interactions adds complexity to the biological system, while also causing concerns for cellular toxicity. Here, we show that the stability of DNA-SWCNTs after internalization by mammalian cells is dependent upon the oligonucleotide sequence that wraps the nanotube. The relationship between sequence length, fluorescence stability, and uptake are examined to provide insight on intracellular mechanisms that require considerations in nanotube sensor design. We show that aggregation not only alters the expected rate of uptake into the cells but further affects the endosomal processing pathways. We propose a model for the uptake and expulsion of SWCNTs that accounts for the aggregation state and DNA wrapping. These findings provide fundamental understanding of the interactions between nanotubes and live cells which can be applied towards development of carefully engineered carbon nanotube sensors.

574243 Light Absorption from Gold Nanoparticle Agglomerates

Gold nanoparticles have shown significant potential for biomedical applications, such as in-vivo photothermal treatment of cancer [1]. This is due to their light absorption enhanced by collective oscillations of their surface electrons during interaction with light [2]. These oscillations are efficiently converted to heat, killing cancer cells that may surround gold nanoparticles [3]. Since the electromagnetic wave transmittance through human skin reaches a maximum at near-infrared (NIR) wavelengths, it is crucial to optimize the light absorption of gold nanoparticles in this spectrum [4]. The light absorption of single gold spheres reaches a maximum at a wavelength of 530 nm, i.e. far away from the NIR spectrum. Non-spherical gold nanoparticles, such nanorods [5], have enhanced NIR light absorption. Similarly, agglomerates of gold nanoparticles formed by coagulation have enhanced NIR absorption due to plasmonic coupling effects between their constituent primary particles [1].

Here, discrete element modeling (DEM) is coupled with discrete dipole approximation (DDA) [6] to investigate numerically the evolution of gold light absorption during nanoparticle agglomeration. The DDA model is validated against simulations and experiments for single spheres and nanorods [5]. The evolution of agglomerate size distribution and morphology are derived by a DEM for agglomeration of gold primary particles. The dynamics of gold light absorption derived by DEM-DDA are in good agreement with those measured from gold nanoparticle agglomerates. The gold nanoparticle agglomerate morphology and size distribution with the optimal NIR light absorption are determined and benchmarked against those of single spheres and nanorods [5].

References:

[1] Sotiriou, G. A., Starsich, F., Dasargyri, A., Wurnig, M. C., Krumeich, F., Boss, A., Leroux, J. C., & Pratsinis, S. E. (2014) Adv. Funct. Mater. 24, 2818-2827.

[2] Willets, K. A., & Van Duyne, R. P. (2007) Annu. Rev. Phys. Chem. 58, 267-297.

[3] Radt, B., Smith, T. A., & Caruso, F. (2004) Adv. Mater. 16, 2184-2189.

[4] Weissleder, R. (2001) Nat. Biotechnol. 19, 316-317.

[5] Qin, Z., Wang, Y., Randrianalisoa, J., Raeesi, V., Chan, W.C.W., Lipinski, W., & Bishof, J.C. (2016) Sci. Rep. 6, 29836-1-13.

[6] Kelesidis, G. A., & Pratsinis, S. E. (2019). Proc. Combust. Inst. 37, 1177-1184.

574486 Ion-Induced Nucleation of Carbonaceous Nanoparticles Ion-induced nucleation of carbonaceous nanoparticles

Kimberly Bowal[1], Jacob W Martin[1], Laura Pascazio[1], Markus Kraft[1,2,3]

[1]*University of Cambridge,* [2]*Nanyang Technological University,*

[3]*Cambridge Centre for Advanced Research and Education in Singapore*

Carbonaceous particles, known as soot, are by-products of incomplete combustion that negatively impact combustion devices, human health and the environment. There is a strong desire to mitigate these undesired effects through the use of clean fuels or redesign of combustion devices. This requires detailed knowledge of the complex chemical and physical processes involved in the formation and growth of soot particles, including the gas-to-solid transition of hydrocarbon molecules to form soot particles. This process is poorly understood, largely due to its low species concentrations, short time frame, and small length scales, which limit experimental measurements. Molecular modelling tools are not restricted by these conditions and therefore provide valuable insight into soot nanoparticles. It is known that curved polycyclic aromatic hydrocarbons (cPAHs) possess significant dipole moments and are found within flames [1]. Recent work suggests that long-range interactions between these curved species and chemi-ions may be important to the soot formation mechanism [2].

This work explores the impact of these ion-dipole interactions in stabilising nascent soot particles and works towards the development of a heterogeneous nucleation model for these species. A force field is developed that is able to capture the flexoelectric effect and enhanced electrostatic interactions. This is used within advanced molecular dynamics simulations to provide information on homogeneous and ion-induced heterogeneous nucleation of cPAHs. The clustering behaviour and resulting morphologies are explored across a range of temperatures. These results provide insight into how the interactions between cPAHs and ions affect the formation and structure of soot nanoparticles and provide valuable information towards expanding current soot models.

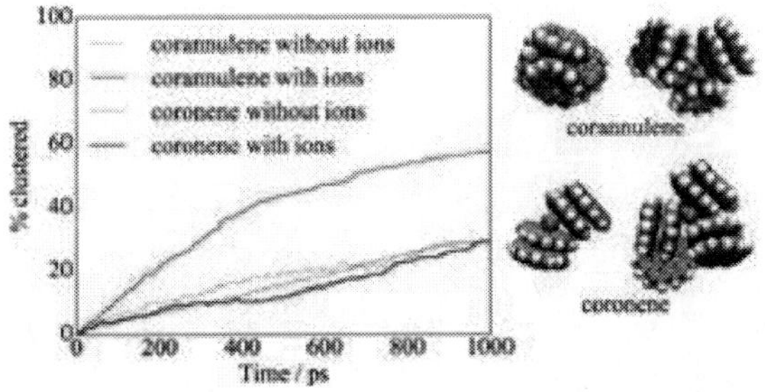

Figure 1: Percent of molecules clustered over time for curved PAH (corannulene) and planar PAH (coronene) systems, with and without potassium ions present. Representative cluster images are shown on the right for corannulene and coronene systems containing ions.

1. Martin, J. W., Slavchov, R. I., Yapp, E. K. Y., Akroyd, J., Mosbach, S., & Kraft, M. (2017). *Journal of Physical Chemistry C*, 121(48), 27154–27163.

2. Martin, J.W., Bowal, K., Menon, A., Slavchov, R.I., Akroyd, J., Mosbach, S., Kraft, M. (2019). *Proceedings of the Combustion Institute*. 37(1), 1117-23.

575202 Nanowire Organic Sensor for Ammonia Detection

Ammonia is one of the largest produced chemicals with a global production of 150 million tons in 2017, expected to increase 8% in the next 4 years. It is widely used in several industries mainly in fertilizers manufacturing, but also in refrigeration systems, production of plastics, and as intermediates for pharmaceuticals synthesis. However, ammonia is highly corrosive and can damage cells in the body in contact, posing serious health risks. Despite the developments made in the detection systems, ammonia is still one of the most dangerous substances, responsible for the second highest number of chemical exposure casualties. The current detection solutions have serious limitations to be broadly used in industrial safety. Detection systems can generally fall under two categories, high-end equipment that is expensive, not user-friendly and can be only used in lab settings, and affordable devices that is prone to false alarms in a varying environment (humidity, temperature, etc.).

In this presentation, we will describe our Nanowire Organic Sensor technology as an alternative solution platform for ammonia monitoring that is affordable and can provide real-time and wide-area coverage. Our patented technology is a solution-based, room-temperature process, which facilitates scalable manufacturing directly on substrates and devices. Compared with the existing nanowire sensor fabrication methods, out innovation has four main unfair advantages. 1) By direct deposit of a solution droplet, nanowires are created right on chips or micro-fabricated devices. 2) It is a room temperature fabrication process, thus enabling less complex and capital-intensive manufacturing process. 3) It is a modular method that enables combinatory synthesis of a wide range of new organic nanowires. 4) It is a fabrication technology compatible with flexible electronics platforms. We will show how the combination of the sensor arrays, impedance spectroscopy and artificial neural networks model enhances the selectivity in varying levels of humidity. We will also describe how the technology is compatible with the internet of things technology and cloud services, and how the combination could be leveraged to ensure more flexible monitoring of ammonia for industrial safety.

575355 Bandgap Engineering of 2D Metal Oxides Using Sulfur for Catalytic and Electronic Applications

As we move towards the future, it will be necessary to develop materials that are earth-abundant, affordable, and environmentally friendly for applications in sectors such as energy and electronics. In this respect, 2D metal oxides are attractive because they can be made of earth-abundant materials and they possess a large surface area that is beneficial for catalysis. In order to modulate the activity of metal oxides towards particular reactions, some researchers have developed and applied doping methods which can alter and/or improve their properties. Sulfur has been applied as a popular dopant for many different catalysts, but there is less research exploring the effect of sulfur-doping on 2D transition metal oxides. In order to begin understanding the influence of sulfur-doping in 2D systems, we have controllably doped CuO and Fe2O3 nanosheets with sulfur and characterized their properties. X-ray photoelectron spectroscopy, diffuse reflectance spectroscopy, X-ray diffraction, electrical resistivity measurements, and heterogeneous catalytic reactions were used to characterize the composition, bandgaps, catalytic activity, and transport properties of CuO and Fe2O3 nanosheets with varying levels of sulfur-doping. The effect of sulfur doping concentration on bandgap, catalytic activity, and transport properties will be discussed, and the influence of sulfur on modulating properties will be compared between CuO and Fe2O3 2D systems.

575770 Synthesis and Separation of Metallic Nanoparticles in Aqueous Two Phase System

In this work, we demonstrate a novel synthesis of silver nanoparticles in an aqueous two-phase system. Aqueous two-phase system (ATPS) is a biphasic system consisting of two aqueous phases. They can be formed by the use of polymer-polymer mixture with different hydrophilicity or a polymer-salt mixture. Here, polyethylene glycol and sodium citrate are used as two-phase forming components. Novel synthesis of silver nanoparticles in ATPS is achieved by use of sodium hydroxide as a catalyst. Addition of sodium hydroxide also allows room temperature synthesis of these nanoparticles. Particles with size 60 nm are obtained in the top phase with polyethylene glycol as a primary reducing agent. The bottom phase consisting of citrate-rich solution yielded nanoparticles with size around 30 nm. Change in the particle morphology in different phases is attributed to different reduction mechanism of both the phases.

576143 Facile Method to Fabricate Nanostructured Silver Selenide Thin Films at Room Temperature from a Copper Selenide Template

Ag_2Se is a promising n-type candidate for thermoelectric applications. Nanostructured Ag_2Se reduces the thermal conductivity and increases Seebeck coefficient due to the grain boundaries between the nanoparticles. In contrast to other complex and expensive methods of synthesizing Ag_2Se, our group reports a facile and efficient strategy of fabricating nanostructures Ag_2Se thin films via ion exchange technique at room temperature. Starting from a cubic-phase $Cu_{2-x}Se$ template prepared by thiol-amine dissolution process, Ag_2Se thin films were created by simply soaking our $Cu_{2-x}Se$ thin films into a Ag^+ ion solution for a short amount of time while maintaining the morphology of $Cu_{2-x}Se$ thin films. The transition from $Cu_{2-x}Se$ matrix to post-soaked Ag_2Se thin films was justified through the phase change and compositions analyzed by X-ray diffraction (XRD) and Energy Dispersive X-Ray Spectroscopy (EDS). This safe and solution-processable reaction yields to Ag_2Se thin films with a power factor of 350. The reaction process is far more economical, facilitating the scale-up to production, compared to other Ag_2Se nanoparticle fabrication processes, such as Schlenk line synthesis and chemical vapor deposition. We aim to improve the power factor via methods such as elemental doping, atomic layer deposition, and energy filtering. Most importantly, this method of template nanostructures provides a framework for exploring a wide range of thermoelectric materials beyond Ag_2Se thin films.

576687 Diffusion of Species in Functionalized Micro/Mesoporous Au@SiO_2 Core-Shell Nanoparticles

Silica-encapsulated gold core-shell nanoparticles (Au@SiO_2) were synthesized via a bottom-up synthesis to catalyze the selective oxidation of benzyl alcohol. The pore size and morphology of the catalyst allows for the unique opportunity to investigate the impacts of silica functionalization on mass transport and surface chemistry discretely. The nanoparticles exhibit a pore distribution with a peak at 27 Å, a size which enhances selectivity via preferential transport of the desired product (i.e. benzaldehyde) relative to larger, undesired products (i.e. benzoic acid/benzyl benzoate). GC-FID analysis revealed the addition of potassium carbonate during the catalytic oxidation of benzyl alcohol increased conversion from 58% to 75% while only decreasing selectivity from 98.5% to 97.7%.

These results suggest that the pore size distribution within the inert silica shell of Au@SiO_2 physically inhibits the formation of undesired products to facilitate the selective oxidation of benzaldehyde despite a basic environment which would drastically reduce selectivity under typical conditions. Tuning the average pore diameter can allow the investigation of mass transport independent from surface chemistry. Diffusion-Order Spectroscopy NMR (DOSY-NMR) was performed on catalysts with varying pore size and functionality, revealing larger pores exhibited both lower diffusion coefficients and lower selectivity.

578571 Mediatorless, Reversible Optical Nanosensor Enabled through Enzymatic Pocket Doping

Single-walled carbon nanotubes (SWCNTs) exhibit intrinsic near-infrared fluorescence that benefits from indefinite photostability and tissue transparency, offering a promising basis for *in vivo* biosensing. Existing SWCNT optical sensors that rely on charge transfer for signal transduction often require exogenous mediators that compromise the stability and biocompatibility of the sensors. This study presents a reversible, mediatorless, near-infrared glucose sensor based on glucose oxidase-wrapped SWCNTs (GOx-SWCNTs). GOx-SWCNTs undergo a selective fluorescence increase in the presence of aldohexoses, with the strongest response toward glucose. When incorporated into a custom-built membrane device, the sensor demonstrates a monotonic increase in initial response rates with increasing glucose concentrations between 3×10^{-3} and 30×10^{-3} M and an apparent Michaelis–Menten constant of $K_{M(app)} \approx 13.9 \times 10^{-3}$ M. A combination of fluorescence, absorption, and Raman spectroscopy measurements suggests a fluorescence enhancement mechanism based on localized enzymatic doping of SWCNT defect sites that does not rely on added mediators. Removal of glucose reverses the doping effects, resulting in full recovery of the fluorescence intensity. The cyclic addition and removal of glucose is shown to successively enhance and recover fluorescence, demonstrating reversibility that serves as a prerequisite for continuous glucose monitoring.

578580 Spinning-Disc Confocal Microscopy in the Second Near-Infrared Window (NIR-II)

The distinctive properties of single-walled carbon nanotubes (SWCNTs) have inspired the development of innovative applications in the field of cell nanobiotechnology. Most studies to date have focused on eukaryotic cells capable of internalizing SWCNTs functionalized with a variety of non-covalent wrappings. However, the effect of SWCNT functionalization on transport across the thick cell wall of prokaryotes remains unexplored. In this study, we explore the uptake of SWCNTs in Gram-negative cyanobacteria and demonstrate selective internalization of SWCNTs decorated with charged protein wrappings. The functionalized SWCNTs are shown to traverse the outer cell wall of both filamentous and unicellular strains of cyanobacteria, independent of whether the strain is naturally competent for DNA uptake, with adsorption and internalization rate constants of $k_{ads} = (9.08 \pm 0.16) \times 10^{-8}\ s^{-1}$ and $k_{in} = (1.466 \pm 0.011) \times 10^{-4}\ s^{-1}$, respectively. A custom-built, spinning disc confocal microscope was used for the first time to directly image near-infrared (NIR) SWCNT fluorescence within cells, revealing a highly inhomogeneous distribution of SWCNTs that is otherwise overlooked using conventional NIR widefield imaging. The nanobionic cells show sustained photosynthetic activity and growth, offering a powerful avenue for engineering photosynthetic organisms with augmented and even inherited nanobionic capabilities.

578943 Accumulative Roll Bonding (ARB)-Processed High-Manganese Twinning Induced Plasticity (TWIP) Steel with Extraordinary Strength and Reasonable Ductility

Accumulative roll bonding (ARB) process at room temperature was used to fabricate a nanostructured twinning-induced plasticity (TWIP) steel. The effect of the number of ARB cycles from 1 to 3 on the microstructure and mechanical properties of Fe-31Mn-3Al-3Si (wt%) TWIP steel was investigated. The results showed that the microstructure of the 1-cycle ARB-processed sample mainly contained the primary and secondary mechanical twins. However, in the microstructure of 2-cycle ARB-processed sample, new configurations of small nanotwins which they are known as the hierarchical nanotwinned (HNT) structures were observed between the secondary mechanical twins. In the microstructure of 3-cycle ARB-processed sample, new subdivided grains with a mean size of about 100 nm were produced. Additionally, the 3-cycle ARB-processed TWIP steel sample exhibited extraordinary strength (about 1.2 GPa) and reasonable ductility (about 15%). The mechanical behavior of ARB-processed TWIP steel was interpreted as the evidence for microstructural features such as the mechanical twins (especially, the HNT structures) and nano/ultrafine grains.

576628 Engineering Ionically-Gelled Chitosan Nanoparticles Containing the Nucleoside Analog, Gemcitabine

Engineering ionically-gelled chitosan nanoparticles containing the nucleoside analog, gemcitabine

Austin Fergusson[1], Richey M. Davis[2, 3]

[1]*Translational Biology, Medicine, and Health Graduate Program*

[2]*Macromolecules Innovation Institute*

[3]*Department of Chemical Engineering*

Virginia Tech, Blacksburg, VA 24061

Gemcitabine (Gemzar) is a potent, FDA-approved cancer therapeutic for the treatment of late-stage pancreatic cancer. Once inside of a cell, gemcitabine inhibits DNA synthesis by terminating DNA strand elongation which leads to rapid cell death. Despite gemcitabine's successful use in treating various cancers, its efficacy is limited by rapid enzymatic degradation and transporter-mediated trafficking into cells. A nanoparticle drug delivery system could protect gemcitabine from degradation and bypass the need for a transporter. Flash Nanoprecipitation (FNP) can produce nanoparticle populations with well-defined sizes and high drug loading capacities for lipophilic therapeutics (Log P = 3.5); for charged, hydrophilic therapeutics, additional charge interactions or some entrapment mechanism is needed. Prior work using a slow nanoprecipitation method resulted in significant loadings of gemcitabine in chitosan nanoparticles when the Pluronics F-68 PEO-PPO-PEO triblock copolymer was present. We hypothesized chitosan coacervation could provide an avenue to encapsulate hydrophilic therapeutics using FNP and produce nanoparticles with high loading capacities of gemcitabine. The pKa value of chitosan, 6.5, also provides a useful pH-sensitive release mechanism for entrapped drugs.

Chitosan nanoparticles containing gemcitabine loaded at > 12 wt% were fabricated using FNP with ionic gelation. We employed a factorial experimental design to examine the FNP formulation parameters that promoted gemcitabine encapsulation. Nanoparticle sizes ranging from ~ 150 – 650 nm were observed. This presentation will focus on the factorial design and optimization of the FNP formulation, synthesis, and characterization of these chitosan nanoparticles including particle size, zeta potential, and gemcitabine loading and release kinetics. The use of FNP reduces the time necessary to achieve a well-defined chitosan nanoparticle population using ionic gelation and simplifies the process of scaling-up nanoparticle production.

568739 The Effects of Scavenger Receptor Class B Type 1 on the Uptake of Both Hard and Soft Nanoparticles

As medicine becomes more advanced, the inclusion of nanoparticles in drug delivery systems will become increasing popular. A question that needs to be answered is the how metabolic state of the patient will affect how the biodistribution of the nanoparticles. In western society, this is often associated with obesity. Obesity causes many shifts in a patient's metabolic states, including a higher ratio of LDLs (bad cholesterol) to HDLs (good cholesterol). HDLs are said to be good cholesterol because they allow cholesterol to be brought back to the liver for clearance. Because the standard model for obesity (ob/ob mice) don't exhibit the same shift in cholesterol ratios as humans, we generated a high cholesterol model by knocking out scavenger receptor class B type 1 (SR-B1). SR-B1 binds HDLs at the liver. SR-B1 has been previously shown to also affect the uptake of HDL coated silver NPs and connected to the macrophage activation marker CD-86. To test if SR-B1 also has the ability to affect the uptake of uncoated HDLs, we conducted *in vitro* tests with hard NPs (polystyrene (PS)) and soft, filomicelle NPs (PBD-b-PEO). The filomicelles were selected because they are known to have weak affinities for plasma proteins. On the other hand, PS-COOH NPs to have sub micromolar affinities for HDLs. Nanoparticle uptake assays were conducted with and without a SR-B1 inhibitor (BLT-1) for a microphage (RAW 264.7) and an epithelial liver (Hep G2) cell line. Regardless of pre-incubation with mouse plasma, BLT-1 treatment resulted in a 3-fold decrease in PS-COOH and filomicelle uptake in RAW. Hep G2 cells didn't uptake a significant amount of either NP in any condition. This shows that SR-B1 may have impacts on nanoparticle-based drug delivery beyond HDL targeting, as previously thought.

575315 Process Intensification for Small Molecule Synthesis Via Recombinant Peptide Templated Palladium and the Elimination of Intermediate Purification

To illustrate the concept of process intensification for small organic molecule synthesis, a project was undertaken to test the catalytic ability of recombinant peptide-templated palladium nanoparticles (Pd NPs) in an impure environment. First, the soluble fraction of bacterial lysates containing repeat units of a metal binding domain (Pd4) fused to the carrier protein Green Fluorescent Protein was used to direct palladium nanoparticle synthesis without purification. Highly ordered NPs on the order of 2nm were prepared and then used for cross-coupling reactions commonly used in the pharmaceutical industry (e.g. the synthesis of the anti-cancer drug Tykerb). Data showed that Pd NPs without any purification could catalyze the synthesis of fifteen representative compounds via Suzuki or Stille cross-coupling reactions, with the purification of the drug analogs the only purification step of importance. Thus, by using a combination of green chemistry to both produce the catalyst particles and product we illustrate a "one pot process" that requires the minimum number of overall purification steps from start to final product.

560087 A Shapelet-Based Method for Analysis of Self-Assembled Surface Imaging

The determination of quantitative structure-property relations is a vital but challenging task for nanostructured materials research due to the presence of large-scale spatially varying patterns resulting from nanoscale processes such as self-assembly and nano-lithography. Focusing on nanostructured surfaces, recent advances have been made in automated quantification methods for translational order using shapelet functions, originally developed for analysis of images of galaxies, as a reduced-basis for surface pattern structure.

In this work, a method combining shapelet functions and a machine learning methods is developed and applied to a representative set of images of self-assembled surfaces from experimental characterization techniques including SEM, AFM, and TEM. The method is unguided, computationally efficient, and able to quantify salient pattern features including deformation, defects, and grain boundaries from a broad range of patterns typical of self-assembly processes.

560096 Reorientation Dynamics and Symmetry-Breaking of Confined Nematic Liquid Crystal Domains

Liquid crystals (LCs) refer to a class of materials which have anisotropic properties. They are used in many technological applications ranging from displays to biological sensors. One example of a category of technologically relevant LC applications is optical functional materials, which include polymer-dispersed liquid crystal (PDLC) films, sometimes referred to as "smart" windows. In these films, non-deformable micron-scale LC droplets are dispersed in a solid polymer matrix. Application of an electric field through the thickness of a PDLC film results in "switching" between a transparent "on" state (field-on) and a translucent "off" state (field-off).

In this work, orientationally-ordered nematic LC phase formation and external electric field-switching dynamics of non-deformable droplets are studied using the continuum Landau-de Gennes model. The model is able to capture phase transition and reorientation dynamics on device-relevant length and time scales when combined with numerical methods such as the finite element method. Formation dynamics correspond to transitioning from a high-temperature disordered liquid phase to an orientationally-ordered phase referred to as a nematic LC. Field-switching dynamics correspond to the imposition and release of an external (electric) field. The effects of non-spherical shape is studied, which may form naturally or intentionally under controlled conditions in the manufacturing of PDLC films. The interactions between shape, LC/polymer interfacial effects, electric field strength, and other parameters are investigated using fully three-dimensional transient simulations. Additionally, simulation results predicting a symmetry-breaking phase transformation process for sub-micron scale spherical droplets are presented.

Metal-Encapsulated Carbon Nanotubes to Enhance Dispersion in Metal-Matrix Composites

Gregory Chester‡, Mainstream Engineering, Rockledge, FL
Anna Skinner, Mainstream Engineering, Rockledge, FL
Kirk Ziegler, University of Florida, Gainesville, FL
Justin Hill, Mainstream Engineering, Rockledge, FL‡
‡Corresponding Author: gchester@mainstream-engr.com

Abstract

Carbon nanotubes (CNTs) remain a topic of great interest since the prediction and discovery of their potential for impressive electrical and thermal properties in the early 1990s. Integrating CNTs into functional composite materials has proven difficult due to their propensity to aggregate and poor interfacial adhesion with the surrounding matrix. Here-in, we will describe a method to reduce CNT aggregation and improve interfacial adhesion with a metal matrix by encapsulating individual or small bundles of CNTs in a metal shell. The metal-encapsulation process takes advantage of the solvent shell surrounding surfactant-suspended CNTs by reducing metal salts at the phase boundary. Using this encapsulation, we have demonstrated a 3x minimum improvement in average axial fatigue life vs. untreated 1080 steel. This process presents a potential pathway to improvements in metal composite physical properties.

Introduction

Since first being reported in 1991,[1] CNTs have been of great interest to the scientific community for their incredible physical, electrical, and thermal properties.[2] Researchers have sought to harness these properties in bulk materials by integrating CNTs into functional polymeric[3] and metallic[4] composite materials. Individual CNTs have a demonstrated Young's modulus up to 3.6 TPa and have theoretical tensile strengths over 100 GPa.[5] This is almost 100x stronger than many steels.[6] Integrating high-strength reinforcing additives into polymers or metals can increase the strength of the resulting composite as a function of additive strength, loading, and alignment. Ceramic reinforcing wires are commonly used in metal matrix composites (MMCs);[7] however, ceramic additive composites can have increased weight and reduced flexural strength. In recent years, CNTs have garnered interest as additives for MMCs because of their tensile strength, fatigue resistance, and electrical and thermal conductivities. Integrating CNTs into useful products remains a challenge because of their tendency to aggregate[8-9] and poor interfacial adhesion between CNTs and the matrix,[3] leading to reduced performance. Although dispersion is a significant challenge for all nanomaterials,[10] CNTs are particularly problematic because of strong Van der Waals interactions.[9, 11-13] Aggregated CNTs form defect spaces in the MMC that inhibit realization of maximum strength improvements.

For effective mixing and chemical processing, CNTs must be suspended in solution, which inhibits aggregation. Common solvents do not provide sufficient solvation forces to suspend individual CNTs.[14] Chlorosulphonic acid is the only solvent that has demonstrated thermodynamically stable solutions; however, safety risks are a concern when working with highly reactive acids.[15] Therefore, CNTs are often dispersed in aqueous suspensions with the aid of surfactants.[16] Most aqueous dispersions use sodium dodecyl sulfate, sodium dodecylbenzene sulfonate (SDBS), and sodium cholate, although other surfactants and polymers have been used.[17-20] After suspending the CNTs, organic solvents can swell the hydrophobic core of surfactant micelles surrounding carbon nanotubes to yield an emulsion-like environment surrounding the CNTs, as shown in Figure 1.[21] Metal salts can be dissolved in the organic solvent and subsequently reduced to create a coating on the individually suspended CNTs. This coating provides benefits similar to functionalization;

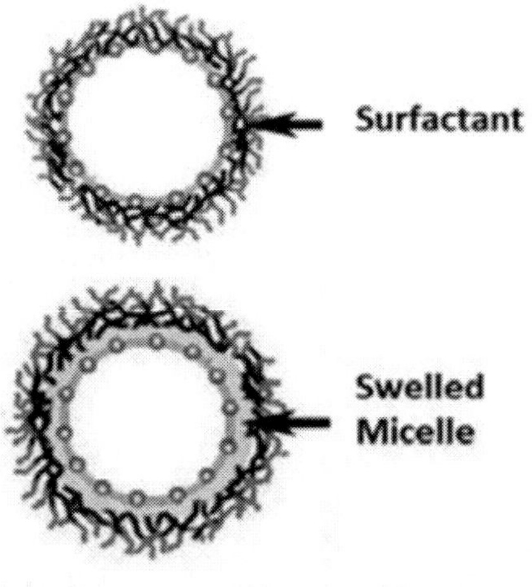

Figure 1: Carbon nanotubes can be individually suspended in an aqueous solution through the addition of a surfactant. The micelle around the CNT can then be swelled by mixing with an organic solvent that stays in the oleophillic shell within the micelle. A metal salt can be dissolved in the organic solvent to keep the metal immediately at the CNT surface.

however, the process does not destroy the covalent bonding in the CNT.

To improve interfacial adhesion and dispersion without reduction in CNT properties, we have demonstrated a method to metal-encapsulate CNTs through use of the solvent shell around a surfactant-suspended CNT. In addition, the synthesis approach can be adapted easily to different coating materials and different dimensions of the nanomaterial such as multi-walled CNTs, single-walled CNTs (SWNTs), or graphene flakes.[22] The key component to the micellar reaction method is a two-phase immiscible system: a surfactant and a precursor, each soluble only in one of the two immiscible phases, and a reducing agent that is soluble only in the other immiscible phase. By using precursors and solvents with these characteristics, the reaction is forced to occur only within the surfactant micelle. This is a common technique for forming nanoparticles.[23-25]

The metal encapsulation approach is shown in greater detail in Figure 2, which shows the coating process on a single CNT. SWNTs are dispersed in water with the aid of a surfactant, such as SDBS, using high-energy homogenization and ultrasonic mixing. The dispersion is then separated via ultracentrifugation to separate the well-suspended CNTs, shown in Figure 2(a), from the remaining aggregates. The micelle around the CNT is swelled by mixing the suspension with an immiscible oil phase or organic solvent such as chloroform containing a metal salt, as shown in Figure 2(b). A reducing agent, soluble only in the aqueous phase, is added to the suspension, as shown in Figure 2(c). The metal salt is reduced at the interface, causing the metal to collapse around and encapsulate the CNT, as shown in Figure 2(d). Finally, the metal-encapsulated CNTs can be collected via filtration or centrifugation, and the remaining surfactant can be rinsed away,

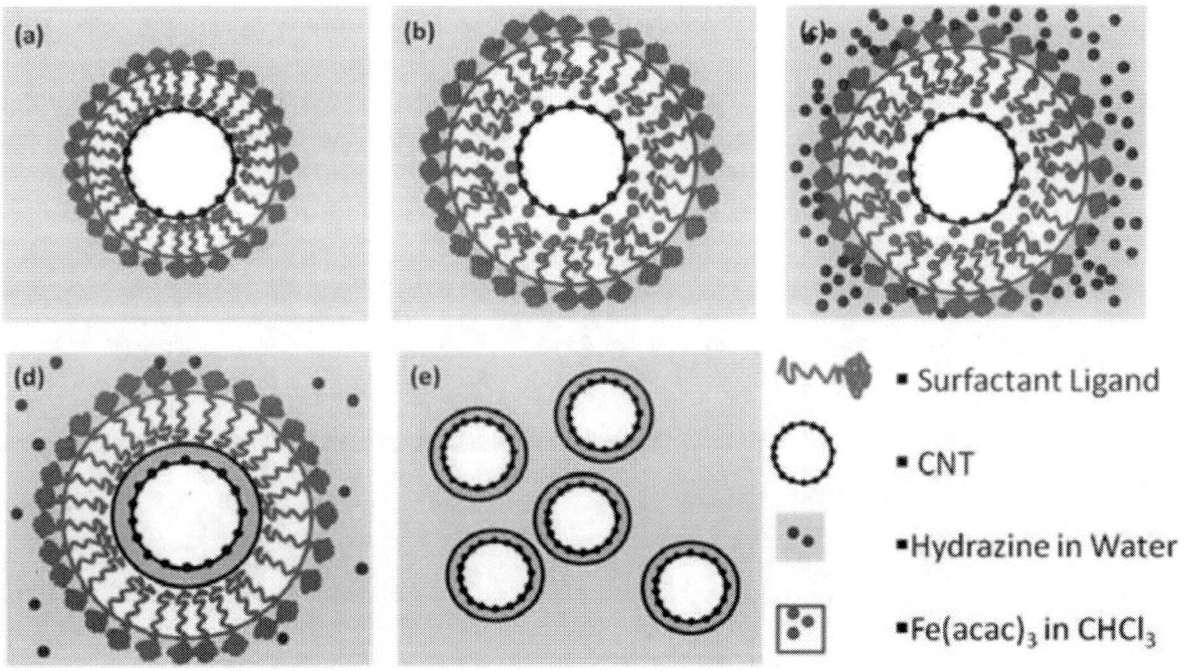

Figure 2: The CNT coating process takes advantage of the solvent shell around surfactant-suspended CNTs to reduce the metal salt directly at the CNT surface.

leaving a powder of individually coated CNTs, as shown in Figure 2(e). The metal-coated CNT powder can be mixed into a metal powder and formed into solid MMCs.

Experimental Procedure

To metal-encapsulate the CNTs, equal volumes of CNT suspension and 0.15 M iron acetylacetonate [Fe(acac)$_3$] in CHCl$_3$ were mixed and allowed to phase-separate for 8 h, which swells the micelles surrounding the suspended CNTs. The portion that contains the suspended CNTs was removed and allowed to rest overnight to evaporate much of the excess chloroform. The following day, 35% N_2H_4 in H_2O was added in a 7:1 molar ratio of N_2H_4 to Fe(acac)$_3$. After this reaction proceeded for 24 h, an equal volume of methanol was added to the solution. The methanol shed the surfactant and allowed the Fe-encapsulated CNTs to settle out of solution. The excess liquid was decanted off, and the remaining Fe-encapsulated CNTs were transferred to centrifuge tubes. These were centrifuged for 2 h at 1000 rpm to force the CNT solids out of any remaining liquid. The supernatant was removed, and the Fe-encapsulated CNTs were transferred to a mixing vessel and dried under vacuum at 30 °C. Samples of dried Fe-encapsulated CNTs were analyzed using scanning electron microscopy (SEM) and energy dispersive x-ray spectroscopy (EDS), which showed an even dispersion of Fe on the CNT surfaces, as shown in Figure 3. The Ni particles in Figure 5: Axial fatigue cycles to failure for plain steel, samples with integrated Fe-encapsulated CNTs, and samples with integrated non-encapsulated CNTs. The percent loadings represent a volume percent loading.

are catalyst remnants from the CNT synthesis, while the O content is due to oxidation of the Fe after encapsulation.

The dry Fe-encapsulated CNT powders were portioned and mixed with 1080 steel powder to create composites with

0%, 0.1%, 0.5%, and 1.0% FeCNTs by volume. Uncoated CNTs were also mixed with steel powder at 0.1% and 0.5% by volume in order to determine the specific benefits related to the encapsulation. These uncoated CNTs were subject to the entire encapsulation process; no $Fe(acac)_3$ was added during the process so that no metal coating would be applied. The CNT/steel powders were mixed using a 2-axis mixer for 30 s total in two 15-s cycles. FeCNT solids were formed by spark plasma sintering (SPS) the steel/Fe-encapsulated CNT powders. The SPS was performed at 80 MPa with a 200 °C/min ramp to 800 °C, which was held for 15 min. Dog-bone shapes were milled using electrical discharge machining for axial fatigue testing.

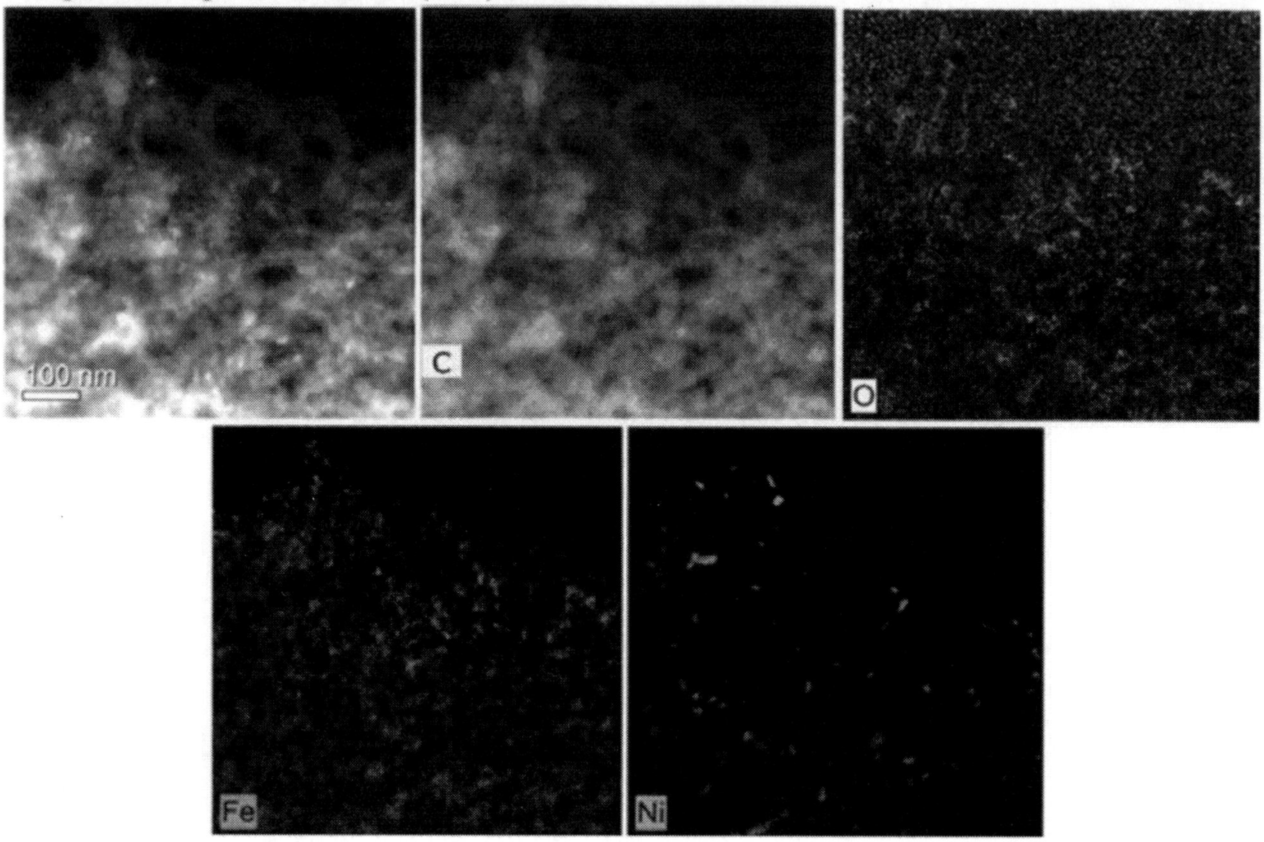

Figure 3: EDS map of CNT cluster showing Fe/FeO_x decoration and Ni particles remaining from CNT synthesis.

The axial fatigue testing was performed in order to show how the addition of Fe-encapsulated CNTs improves the lifespan of a cyclically fatigued material. Tests were performed with a cycling frequency of 30 Hz, stress range of 46.5 ksi, and an r-ratio—the ratio of tension to compression force—of 3. Samples were tested either up to failure or until $5x10^7$ cycles were reached. Only samples that failed under the $5x10^7$ cycle cap were included in the cycle-to-failure data. The number of samples that reached the stop condition were recorded in order to determine if a relationship to Fe-encapsulated CNT addition and/or loading existed.

Results and Analysis

Initial testing with plain 1080 steel showed an average fatigue to failure of 7.2×10^4 cycles, with a low 9×10^3 standard deviation and no samples reaching 5×10^7 cycles. Integrating Fe-encapsulated CNTs into the 1080 steel led to a minimum of 3x improvement in the average fatigue life where 0.1% and 1.0% Fe-encapsulated CNTs had 1×10^5 and 3.4×10^5 cycles to failure, respectively. The 0.5% by volume sample showed 2.4×10^5 cycles to failure. The results are shown in Figure 5. The samples produced using uncoated CNTs in the composite had significantly lower axial fatigue life, which was more in line with the plain 1080 steel. These samples had cycles to failure of 8.3×10^4 and 1.5×10^5 for 0.1% and 0.5%, respectively. While this is an improvement from the plain 1080 steel, it is a significant drop-off from the Fe-encapsulated CNT composite.

The cycles to failure did not appear to show a specific trend with regard to loading Fe-encapsulated CNTs in 1080 steel; a significant standard deviation was associated with Fe-encapsulated CNT composites. The 0.1%, 0.5% and 1.0% composites showed standard deviations of 1.8×10^5, 1.2×10^5, and 1.8×10^5, respectively, which is approximately a 50% error. This error is extremely high, though we believe it to be a function of inconsistent CNT distribution within the composite. The potential for CNT distribution to drive inconsistent fatigue life is further emphasized when accounting for the number of samples that reached the 5×10^7 cycle runout condition. For the Fe-encapsulated CNT composites and the 0.5% uncoated CNT composite sample, 40% – 50% of the samples tested reached the runout condition. For the 0.1% CNT, 80% of samples tested reached the runout condition, as shown in Figure 4.

Many fatigue failures initiate at vacancy or high-strain sites within the metal grain structure. When integrating a reinforcing additive into a metal, the risk of vacancies increases and the strain around the additive sites increases, which can lead to premature failure. For the samples that reached 5×10^7 cycles without failure, we believe the CNTs are better dispersed, which limits agglomerations, reducing vacancies, and minimizes strain within the grain structure.

Conclusions

Integrating CNTs into MMCs is one method of producing very high fatigue life composites. Traditionally processed CNTs often face issues with aggregation and poor interfacial bonding with the metal matrix, which inhibits optimized performance. We have demonstrated a method to encapsulate CNTs in a metal shell by reducing a metal salt in the solvent shell around a surfactant-suspended CNT. These metal-encapsulated CNTs can then be mixed with metal powders and sintered to form MMCs. Axial fatigue life testing showed a minimum 3x improvement for composites containing Fe-encapsulated CNTs. Testing with uncoated CNTs showed a lesser improvement in axial fatigue life, demonstrating the benefits to Fe-encapsulated CNTs. A significant amount of variability was found within a single-composite composition that was likely due to poor mixing of the CNTs in the composite, leading to vacancy sites and strain in the grain structure. The variability is further demonstrated by the number of samples (>40% of samples tested) that reached the runout condition. Moving forward, we will investigate methods to improve mixing consistency in order to reduce sample variability. We will also look to develop a full S-N curve in order to compare more directly composite performance to alternative materials.

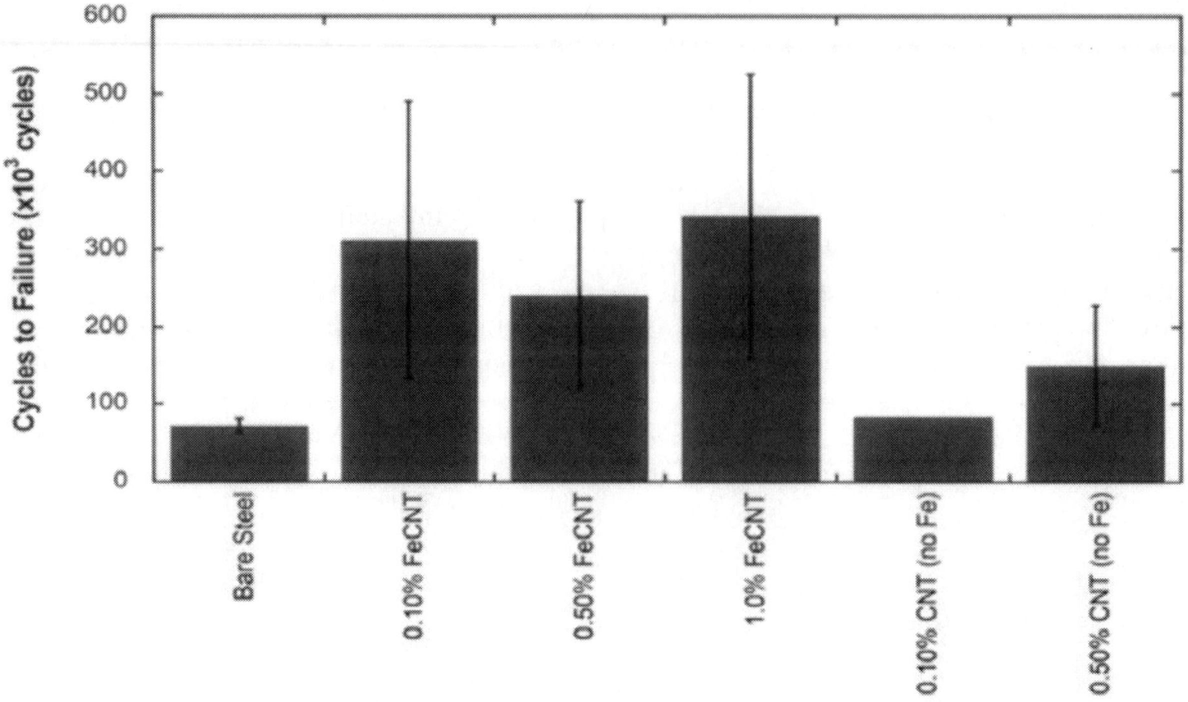

Figure 5: Axial fatigue cycles to failure for plain steel, samples with integrated Fe-encapsulated CNTs, and samples with integrated non-encapsulated CNTs. The percent loadings represent a volume percent loading.

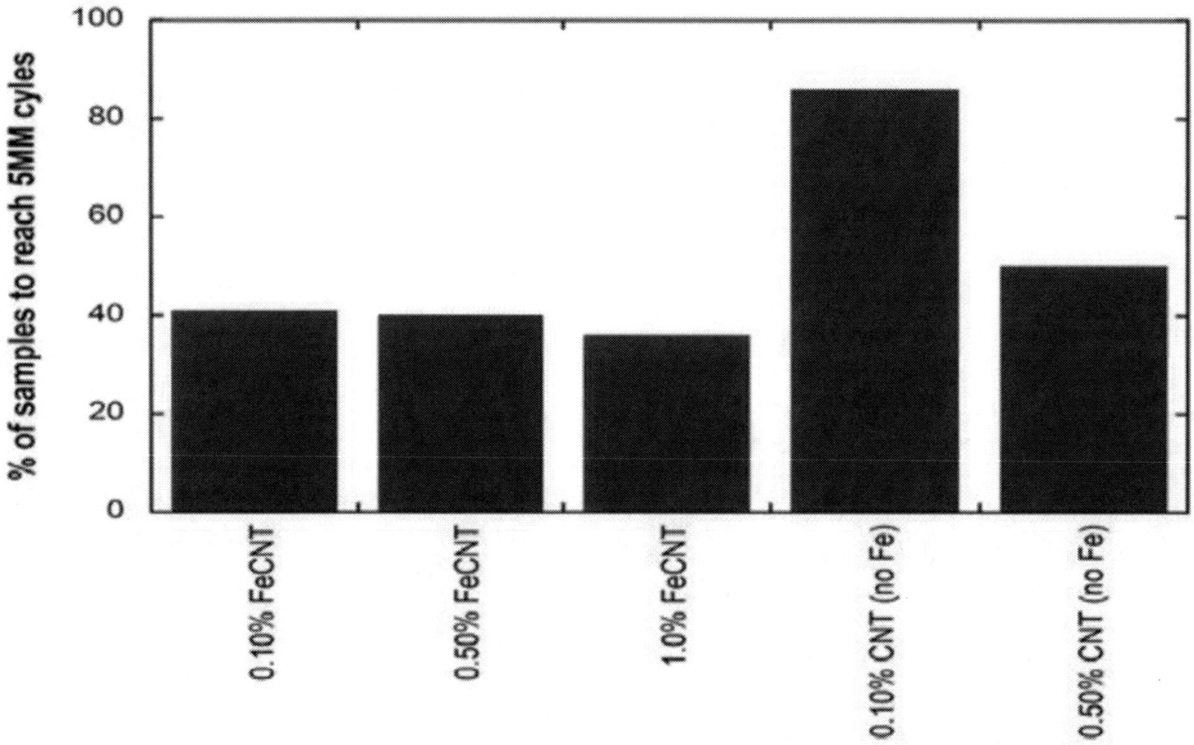

Figure 4: Percentage of each composite type tested that reached $5x10^7$ cycles without failure. The percent loadings represent a volume percent loading.

References

1. Iijima, S., Helical microtubules of graphitic carbon. *Nature* **1991,** *354* (6348), 56-58.
2. Bellucci, S., Carbon nanotubes: physics and applications. *physica status solidi (c)* **2005,** *2* (1), 34-47.
3. Andrews, R.; Weisenberger, M., Carbon nanotube polymer composites. *Current Opinion in Solid State and Materials Science* **2004,** *8* (1), 31-37.
4. Wiegand, D.; Owens, F.; Goyal, A.; Iqbal, Z. *Fabrication of high strength metal-carbon nanotube composites*; Army Armament Research Development and Engeering Center Picatinny Aresenal NJ: 2008.
5. Ruoff, R. S.; Qian, D.; Liu, W. K., Mechanical properties of carbon nanotubes: theoretical predictions and experimental measurements. *Comptes Rendus Physique* **2003,** *4* (9), 993-1008.
6. Gibson, P.; Cress, H. *Analytical study of aircraft arresting gear cable design*; Battelle Memorial Inst Columbus, OH 1965.
7. Suarez, S.; Reinert, L.; Mucklich, F., Carbon Nanotube (CNT)-Reinforced Metal Matrix Bulk Composites: Manufacturing and Evaluation. In *Diamond and Carbon Composites and Nanocomposites*, InTech.
8. Shvartzman-Cohen, R.; Nativ-Roth, E.; Baskaran, E.; Levi-Kalisman, Y.; Szleifer, I.; Yerushalmi-Rozen, R., Selective dispersion of single-walled carbon nanotubes in the presence of polymers: the role of molecular and colloidal length scales. *J. Am. Chem. Soc.* **2004,** *126* (45), 14850.
9. Ajayan, P. M.; Tour, J. M., Nanotube Composites. *Nature* **2007,** *447*, 1066-1068.
10. Mackay, M. E.; Tuteja, A.; Duxbury, P. M.; Hawker, C. J.; Van Horn, B.; Guan, Z.; Chen, G.; Krishman, R. S., General strategies for nanoparticle dispersion. *Science* **2006,** *311*, 1740-1743.
11. Shvartzman-Cohen, R.; Levi-Kalisman, Y.; Nativ-Roth, E.; Yerushalmi-Rozen, R., Generic approach for dispersing single-walled carbon nanotubes: The strength of a weak interaction. *Langmuir* **2004,** *20* (15), 6085-6088.
12. Shvartzman-Cohen, R.; Nativ-Roth, E.; Baskaran, E.; Levi-Kalisman, Y.; Szleifer, I.; Yerushalmi-Rozen, R., Selective dispersion of single-walled carbon nanotubes in the presence of polymers: the role of molecular and colloidal length scales. *J. Am. Chem. Soc.* **2004,** *126*, 14850-14857.
13. Girifalco, L. A.; Hodak, M.; Lee, R. S., Carbon nanotubes, buckyballs, ropes, and a universal graphitic potential. *Phys. Rev. B* **2000,** *62* (19), 13104-13110.
14. Bahr, J. L.; Mickelson, E. T.; Bronikowski, M. J.; Smalley, R. E.; Tour, J. M., Dissolution of small diameter single-wall carbon nanotubes in organic solvents? *Chem Commun* **2001,** (2), 193-194.
15. Davis, V. A.; Parra-Vasquez, A. N. G.; Green, M. J.; Rai, P. K.; Behabtu, N.; Prieto, V.; Booker, R. D.; Schmidt, J.; Kesselman, E.; Zhou, W.; Fan, H.; Adams, W. W.; Hauge, R. H.; Fischer, J. E.; Cohen, Y.; Talmon, Y.; Smalley, R. E.; Pasquali, M., True solutions of single-walled carbon nanotubes for assembly into macroscopic materials. *Nat Nanotechnol* **2009,** *4* (12), 830-834.
16. O'Connell, M. J.; Bachilo, S. M.; Huffman, C. B.; Moore, V. C.; Strano, M. S.; Haroz, E. H.; Rialon, K. L.; Boul, P. J.; Noon, W. H.; Kittrell, C.; Ma, J. P.; Hauge, R. H.; Weisman, R. B.; Smalley, R. E., Band gap fluorescence from individual single-walled carbon nanotubes. *Science* **2002,** *297* (5581), 593.
17. Bandyopadhyaya, R.; Nativ-Roth, E.; Regev, O.; Yerushalmi-Rozen, R., Stabilization of Individual Carbon Nanotubes in Aqueous Solutions. *Nano Lett.* **2002,** *2* (1), 25.
18. Haggenmueller, R.; Rahatekar, S. S.; Fagan, J. A.; Chun, J. H.; Becker, M. L.; Naik, R. R.; Krauss, T.; Carlson, L.; Kadla, J. F.; Trulove, P. C.; Fox, D. F.; DeLong, H. C.; Fang, Z. C.; Kelley, S. O.; Gilman, J. W., Comparison of the quality of aqueous dispersions of single wall carbon nanotubes using surfactants and biomolecules. *Langmuir* **2008,** *24* (9), 5070.
19. Moore, V. C.; Strano, M. S.; Haroz, E. H.; Hauge, R. H.; Smalley, R. E.; Schmidt, J.; Talmon, Y., Individually Suspended Single-Walled Carbon Nanotubes in Various Surfactants. *Nano Lett.* **2003,** *3* (10), 1379.

20. Green, A. A.; Hersam, M. C., Solution Phase Production of Graphene with Controlled Thickness via Density Differentiation. *Nano Letters* **2009,** *9* (12), 4031-4036.
21. Wang, R. K.; Chen, W.-C.; Campos, D. K.; Ziegler, K. J., Swelling the Micelle Core Surrounding Single-Walled Carbon Nanotubes with Water-Immiscible Organic Solvents. *J. Am. Chem. Soc.* **2008,** *130* (48), 16330.
22. Das, S.; Wajid, A. S.; Shelburne, J. L.; Liao, Y.-C.; Green, M. J., Localized In situ Polymerization on Graphene Surfaces for Stabilized Graphene Dispersions. *ACS Appl. Mater. Interfaces* **2011,** *3* (6), 1844-1851.
23. Pileni, M. P.; Lisiecki, I., Nanometer metallic copper particle synthesis in reverse micelles. *Colloids and Surfaces A: Physicochemical and Engineering Aspects* **1993,** *80* (1), 63-68.
24. Brege, J. J.; Hamilton, C. E.; Crouse, C. A.; Barron, A. R., Ultrasmall copper nanoparticles from a hydrophobically immobilized surfactant template. *Nano Lett* **2009,** *9* (6), 2239-42.
25. Threads and Their Strength. In *Introduction to the Design and Behavior of Bolted Joints, Fourth Edition*, CRC Press: 2007.

577636 Reengineering the Tumor Microenvironment to Enhance Nano-Immunotherapy

Nanotherapeutics have been approved for cancer patients, but their effect on survival has been limited and often less than other approved drugs. Immunotherapy has revolutionized treatment of multiple malignancies. Unfortunately, most patients do not benefit from the current immunotherapies and many encounter immune-related adverse effects. In contrast, nanomedicine can reduce some toxicities. The abnormal tumor microenvironment (TME) is a major impediment to the success of both nano- and immune-therapies. Specifically, the barriers posed by the TME not only limit drug delivery but also limit their efficacy even when they accumulate in tumors. The recent approval in breast cancer patients of nab-paclitaxel with an immune-checkpoint blocker demonstrates the potential of combining nanomedicine and immunotherapy. Here, I present strategies to harness the potential of nanotechnology-based immunotherapy by reengineering the abnormal TME.

582328 Advanced Removal / Extraction of the Hydrocarbons by Emulsion Liquid Membrane Enhanced by Nanoparticles

The removal/extraction of the hydrocarbons from industrial wastewater and water has been enhanced by Emulsion Liquid Membrane (ELM) combined with nanoparticles. The ELM consists of organic membrane phase (O) that contains tiny drops of the stripping aqueous (internal) phase (W1) dispersed in the wastewater (external) phase (W2) where the pollutants transfer through phase (O) from phase (W2) to phase (W1) to react with the stripping agent. The invention is the use of the nanoparticles in the organic phase (O) which significantly enhances the removal at a shorter time and enhance the emulsion stability. Magnetic nanoparticles Fe2O3 have been selected in order to recover and recycle them by applying a magnetic field at the completion of the separation. This is needed when the process is scaled up and implemented industrially. Furthermore, collecting the magnetic nanoparticles at the end will enhance the step of demulsifying.

Adding 0.1% (W/W) of the Fe2O3 nanoparticles in organic phase (O) yields enhanced removal of 98.05% at 6 min. Also nanoparticles with larger size nanoparticles can be used which add more economic value to the invention for the industrial development of this new wastewater/water treatment process. Enhancing the separation of hydrocarbons by combining magnetic nanoparticles for emulsion liquid membrane is an advance emerging technology that finds wide commercial applications.

580230 Workshop--Part 1 "I Followed the Recipe, Where Is My Cake?": Important Questions to Ask in Considering Transition to Management

You are an individual contributor, and you aspire to become a leader. But have you thought about why you want to change to management and how one makes that transition? Most people deliberate over it for a long time and work towards this goal. The roles you take in the years before you want to make the jump will demonstrate that you are ready for management. This session will focus on steps to take and opportunities to target to prepare for the transition into management.

580231 Workshop--Part 2: I Know the Why. Now What?

If you know the why, the next thing on the list is to figure out what needs to be done to show that you are ready to be a manager. What should you be prepared for? Unfortunately, they don't teach us how to be a manager in school, and degrees don't necessarily land us a management job. In this workshop, we will list down some activities and opportunities that we can take to be able to demonstrate that you are prepared for management roles.

580232 General Discussion

The workshop leaders will summarize and open the session for general discussion.

572846 Invited: Sensor Applications in Bioengineering: Detection of Small Molecules within Complex and Three-Dimensional Biological Milieu

Bioengineering applications in regenerative medicine and single-cell analysis present challenging microenvironments for diagnostic and real-time detection. We have developed and applied sensors to detect small molecules in neuronal, cardiovascular, blood and biofilm systems. In particular, we have aimed to establish easy-to-use sensor platforms that do not interfere with biological processes. These sensors are reagentless and can be deployed in the presence of complex macromolecular solutions that readily foul, degrade or block many traditional sensing approaches. Herein, we describe our efforts towards detection of neuroactive molecules in the brain, hypoxia gradients in three-dimensional cellular scaffolds, and subcellular oxygen concentrations during electrophysiological and chemo-mechanical single cell experiments.

575136 Tissue-like Neurochemical Sensors Operating in Brain and Gut

Normal neurotransmission relies on a precise and delicate balance of chemical signaling mediated by the release of variety neurotransmitters. Among them, the monoamines including dopamine and serotonin are uniquely involved in the regulation of cognitive processes such as emotion, arousal and memory. Therefore, it is of great interest in both basic research and clinical studies to monitor the fluctuation of monoamine in both brain and peripheral nervous systems. As a very soft and complex biomaterial, the human brain is under constant motion and deformation because of different physiological dynamics, while GI tract is a series of soft organs with inherent motility, joined in a long and twisting tube from the mouth to the anus. As a result, rigid neurochemical sensors can cause inflammation to the brain tissue, and neurochemical sensing in awake animals have historically been impossible to traditional rigid and fragile devices.

Softness and stretchability in new-generation neuroengineering tools are essential to minimize the perturbation of motions inherent in biological systems and maximize the physiological fidelity. Here we develop a new neuroengineering platform, based on soft materials, that could allow access to the brain and entire enteric nervous system for neurochemical sensing. The new neurochemical sensor is based on graphene-elastomer composite, which offers chronic, sensitive, and selective sensing of dopamine and serotonin in the brain. Ex vivo studies shows that the soft sensor could accommodate the peristalsis movement of the intestine and continuously monitor the serotonin dynamics via thermal treatment. In vivo experiments also show that the sensor could be implanted in the colon of mice to examine the drug-stimulated serotonin response. The new neuroengineering platform described here can provide unprecedented insights on the microbiota–gut–brain communication in the context of mental disorders and therapeutic interventions, and can lead to new therapies for psychiatric disorders through GI tract without invasive surgery to the brain.

570825 Cell-Based Electrochemical Sensors for Environmental Pollutant Detection

Endocrine disrupting compounds are ubiquitous; they are found in many products intended for common use, including pesticides, plastics, and even some medications. Yet, these chemicals are implicated in health problems ranging from diabetes to cancer. We have developed a cell-based electrochemical strategy to rapidly and sensitively quantify these compounds. Our strategy measures the biological activity of chemically dissimilar xenoestrogens by harnessing the specificity of a human estrogen receptor. This receptor is expressed on the surface of *E. coli* as a component of an electrochemical sandwich assay. Our sensors require no specialized skills to implement and have enabled very sensitive detection (sub-ppb levels) of environmental pollutants. We have successfully detected compounds in from complex solutions, such as baby formula, and have measured estrogenic activity from unknown compounds leeched from a BPA-free baby bottle.

577305 Biomimetic Composites As Microfabricated, Flexible, Degradable Electrochemical Sensors

Biomimetic composites as microfabricated, flexible, degradable electrochemical sensors

- M. Xu, S. Pradhan, R.K. Pal, V.K. Yadavalli

Department of Chemical and Life Science Engineering, Virginia Commonwealth University, Richmond, VA

Biomimetic composites of naturally derived and synthetic polymers provide exciting opportunities to develop mechanically flexible, physiologically compliant, and degradable bioelectronics systems. In prior work, we have shown that naturally derived silk proteins and conducting polymers (specifically poly(3,4-ethylenedioxythiophene) :poly(styrene sulfonate) (PEDOT:PSS)) can form functional bioinks for sensing and other electrochemical applications. These inks combine the mechanical strength, biocompatibility and degradability of silk proteins, with the conductivity and chemical functionalization of the PEDOT:PSS. The fabrication and characterization of photopatternable, water-based conductive inks comprising PEDOT:PSS and synthesized photoreactive silk proteins was earlier shown [1] The presence of photoreactive groups permits a fully aqueous photolithographic strategy to form conductive microelectrodes on both rigid substrates as well as flexible silk films [2].

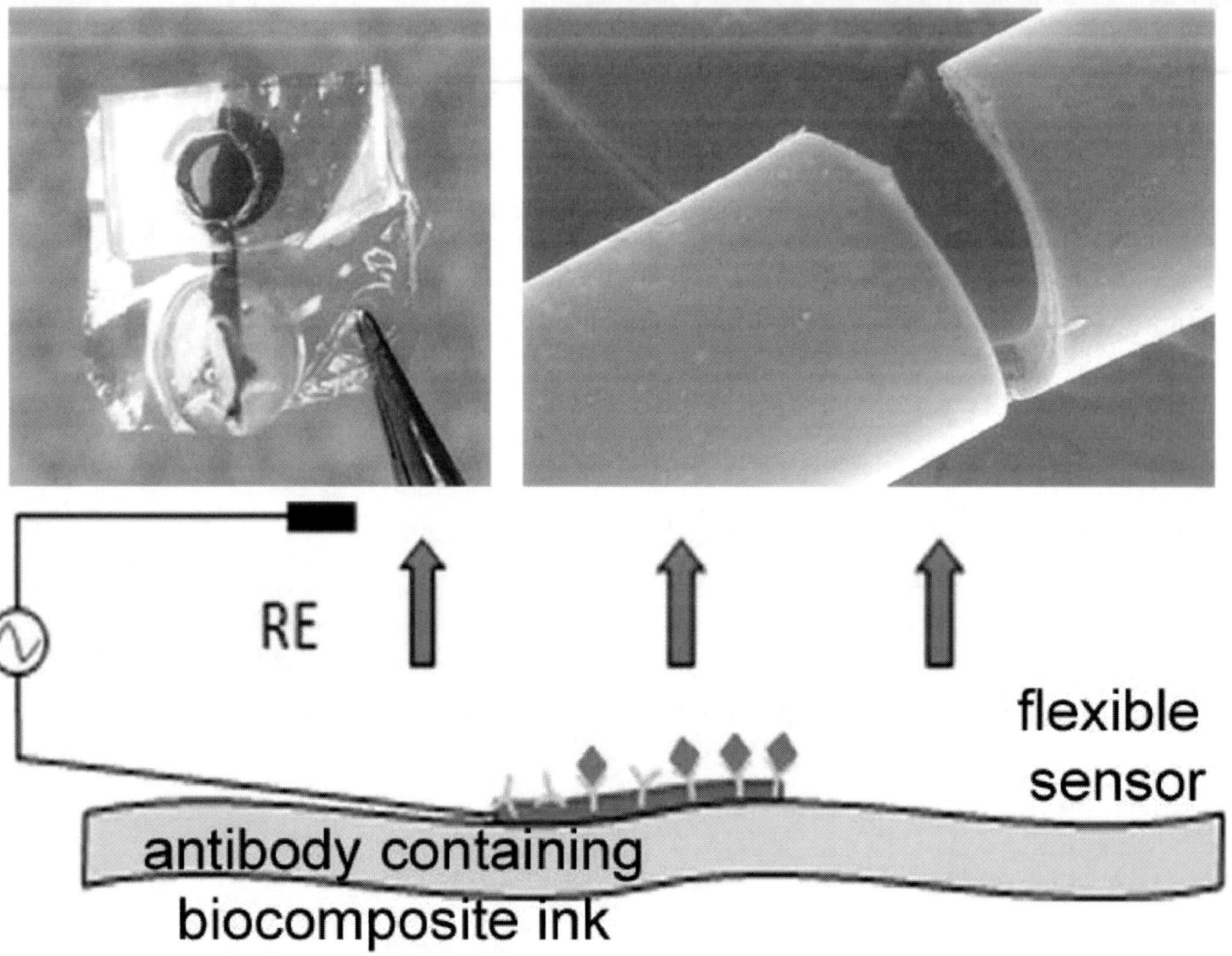

Here we will present the latest developments using this technology, showing how these devices can function for flexible bioelectronics applications in the form of microelectrodes for ultrasensitive detection, as well as components such as microsupercapacitors.[3, 4] We show how electroactive biomolecules such as neurotransmitters can be detected sensitively, while non-electroactive biomolecules such as glucose and glutamic acid, can be detected by encapsulating specific enzymes. Due to the room temperature, aqueous fabrication process, we can immobilize antibodies in these sensors resulting in flexible immunosensors for targets such as vascular endothelial growth factor (VEGF) for the first time [3]. The electroactivity of conductive ink can be improved by the addition of small amounts dopants. Using these doped composites, we further demonstrate flexible energy storage devices due to the capacitive nature of the biomaterial. The presence of silk proteins as the matrix of the composite makes it completely biodegradable and biocompatible, potentially resulting in transient or implantable devices. This platform can also be used for the specific control of cell culture in flexible sheets. The mechanical, biochemical and electrochemical characterization of the composite and its microfabrication are discussed. By virtue of a range of versatile properties, utility as bio-sensors, opto-electronic devices and flexible energy storage systems are envisioned.

References:

[1] "Photolithographic micropatterning of conducting polymers on flexible silk matrices" - RK Pal, AA Farghaly, MM Collinson, SC Kundu, VK Yadavalli, *Advanced Materials*, 28(7), 1406-1412, 2016

[2] "Conducting polymer-silk biocomposites for flexible and biodegradable electrochemical sensors" - RK Pal, AA Farghaly, MM Collinson, SC Kundu, VK Yadavalli, *Biosensors & Bioelectronics*, 81, 294-302, 2016

[3] "Flexible biosensors for the impedimetric detection of protein targets using silk-conductive polymer biocomposites" – M Xu, VK Yadavalli, *ACS Sensors*, in press, 2019

[4] "Fabrication of Flexible, Fully Organic, Degradable Energy Storage Devices Using Silk Proteins" - RK Pal, SC Kundu, VK Yadavalli, *ACS Applied Materials & Interfaces* 10 (11), 9620-9628, 2018

569635 Nanopore Based Highly Sensitive and Accurate Detection of HIV

Nanopore Based High Performance Point-of-Care Biosensor for Acute HIV Detection

Xiaojun Wei, Chang Liu

Human immunodeficiency virus (HIV) is the causative agent of acquired immuno-deficiency syndrome (AIDS). There are over 1.8 million infections HIV occur annually worldwide, and 30%¨C50% are attributed to recently infected individuals that are unaware of their disease state. Rapid diagnosis of acute HIV (within 3 weeks of infection) may reduce new cases by 10%¨C25%. The fourth-generation rapid diagnostic test for HIV is currently employed worldwide as the frontline screening test. This Enzyme-linked Immunosorbent Assay (ELISA) based test is designed to detect HIV-1 p24 antigen in addition to HIV antibodies to achieve early detection. However, several recent field tests have indicated unacceptably low diagnostic sensitivity (0%) of acute HIV-1.

Herein, a nanopore biosensing platform with single-molecule level sensing and analysis capacity has been developed to address the unmet need for rapid and accurate diagnosis of acute HIV to reduce transmission. In our approach, p24 antigens isolated from plasma samples were cleaved into several specific peptide sequences, which can be recognized and quantified by their signature signals during translocation through the nanopore. A portable QuantiPep-HIV prototype suitable for point-of-care settings was developed and tested with serum samples from HIV-infected and uninfected individuals to demonstrate the feasibility and to evaluate the diagnostic performance. Our results showed that QuantiPep-HIV can improve the diagnostic sensitivity and specificity of acute HIV.

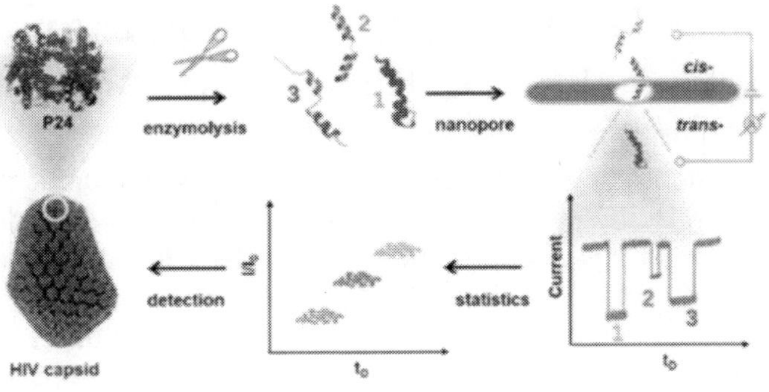

Figure. Schematic diagram for detecting peptide fragments from p24 antigens by nanopore

582378 Wireless, Soft, Skin-integrated Microfluidic Systems for Electrochemical/Colorimetric Biosensing and Capture of Sweat

Human eccrine sweat is a rich composition of nearly 60 known biochemicals including metabolites, electrolytes, vitamins, amino acids, exogeneous drugs, and small proteins. Advances in sensing technologies, conformal electronics, and materials science lay the foundation for recent demonstrations of wearable sweat sensors. Unfortunately, majority of the devices rely either on active, battery-powered electronics for electrochemical detection or passive, colorimetric chemistries for visual readout. Complex construction, large size/weight and possibility of battery leakage represent disadvantages of the former; semi-quantitative operation and limited scope of measurable biomarkers are drawbacks of the latter. Additionally, present wearable sensors can detect only a handful of biomarkers present in the sweat. Quantification of the remaining chemicals still requires sweat collection using adsorbent pads followed by off-site chemical analysis. Such a route provides average concentration of sweat analytes and restricts the acquisition of their temporal concentrations. This prohibits the understanding of the relation between dynamic variations in sweat chemistry and human physiology. In this talk, I will describe an unusual collection of enabling materials, sensing technologies, advanced microfluidics, stimulus dependent biocompatible power systems, galvanic time-tracking schemes, and wireless electronics to address these issues. Specifically, I will describe my strategies that include soft microfluidics, fuel cells, trigger-activated biocompatible batteries, colorimetrics and near field communication technologies that result in new classes of wireless, skin-integrated devices for detailed analysis of human sweat.

574301 The Electrochemical Impedance Response of a Continuous Glucose Monitor

The Gerischer impedance has been observed in systems with an electrochemical reaction coupled with a homogeneous chemical reaction. Under the assumptions of linear homogeneous reaction, equal diffusion coefficients and a semi-infinite condition, the Gerischer impedance can be solved analytically [1, 2]. However, with non-ideal electrochemical systems, where coupled nonlinear homogeneous reactions and various mass-transfer properties make exact analytical resolution impossible, a more detailed analysis is required.

In the present work, a mathematical model for the impedance response of glucose-oxidase based electrochemical biosensors has been developed [3]. The complicated coupling between thirteen homogeneous reactions and three heterogeneous reactions has been considered in the model. The homogeneous reactions included anomerization between α-D-glucose and β-D-glucose and four reversible enzymatic catalytic reactions transforming β-D-glucose and oxygen into gluconic acid and hydrogen peroxide, pH-dependent enzymatic activity and a biological buffer system. The electroactive hydrogen peroxide was considered to be reversibly oxidized or reduced at the electrode.

The mathematical model is solved numerically by using the finite-difference method and Newman's BAND algorithm [4]. The model demonstrates how the coupled non-linear homogeneous reactions effect the diffusion impedance, which has broadened the scope of the Gerischer impedance. The model can be used to explore the influence of various system parameters on limiting current, reaction profiles, and diffusion impedance. The system parameters, including interstitial glucose concentration, oxygen concentration, active enzyme concentration, diffusion coefficients, reaction rate constants and layer thickness, are related to various sensor working conditions such as body sugar level, inflammation, sensor degradation and sensor design.

References

1. H. Gerischer, "Wechselstrompolarisation Von Elektroden Mit Einem Potentialbes- timmenden Schritt Beim Gleichgewichtspotential," *Zeitschrift fur Physikalische Chemie*, **198** (1951) 286–313.
2. M. E. Orazem and B. Tribollet, *Electrochemical Impedance Spectroscopy* (John Wiley & Sons, Hoboken, NJ, 2017), 2nd edition, p. 279-293.
3. M. Gao, M. S. Hazelbaker, R. Kong, and M. E. Orazem, "Mathematical Model for the Electrochemical Impedance Response of a Continuous Glucose Monitor," *Electrochimica Acta*, **275**(2018),119-132
4. J. S. Newman and K. E. Thomas-Alyea, *Electrochemical Systems* (John Wiley & Sons, Hoboken, NJ, 2004), 3rd edition.

Acknowledgement

The support of Medtronic Diabetes (Northridge, CA) and Andrea Varsavsky, program monitor, is gratefully acknowledged.

561693 Collecting Interstitial Fluid from the Skin As a Novel Source of Biomarkers

Tissue interstitial fluid (ISF) is a novel source of biomarkers that complements conventional sources like blood and urine. However, ISF has received limited attention due largely to lack of simple collection methods. To address this issue, we have developed methods to sample ISF from skin in a relatively simple manner with minimally invasive microneedles in animals and human participants.

In one study, we used a microneedle patch to create an array of micropores in skin coupled with mild suction to sample ISF from human participants and identified valuable and sometimes unique biomarkers in ISF when compared to companion plasma samples based on mass spectrometry analysis. Many biomarkers used in research and current clinical practice were common to ISF and plasma. Because ISF does not clot, these biomarkers could be continuously monitored in ISF like current continuous glucose monitors, but without an indwelling subcutaneous sensor. Biomarkers unique to ISF included molecules associated with systemic and dermatological physiology as well as exogenous compounds from environmental exposures. We also determined that pharmacokinetics of caffeine (a model biomarker in healthy adults) and pharmacodynamics of glucose (in diabetic children and young adults) were similar in ISF and plasma, indicating that dynamic information of biomarkers could also be captured using ISF.

In another study, we developed a different microneedle-based method to collect ISF from the skin by pressing microneedles at an angle just below the skin surface to collect blood-free ISF through micropores to the skin surface without the need for vacuum, which was absorbed into a thin strip of paper on the microneedle patch backing for subsequent analysis. An optimized method in rat skin in vivo was well tolerated and able to collect microliters of ISF within minutes. Brief skin pre-treatment with microneedles followed by a delay for a few minutes dramatically increased subsequent ISF collection by a mechanism believed to involve increased skin hydration. Using this approach, we demonstrated the use of ISF for therapeutic drug monitoring by showing similar vancomycin pharmacokinetic profiles in ISF and serum from rats. We also measured polio-specific neutralizing antibodies and anti-polio IgG in ISF similar to serum in rats immunized with polio vaccine.

Overall, these studies provide a method to sample ISF in a relatively simple manner with minimally invasive microneedles and demonstrate ISF as a source of biomarkers that can enable further research and translation for future clinical applications.

569506 From Drug Self-Assembly to Supramolecular Medicine

Drugs are a special class of chemical substances that produce a biological effect when administered to a living organism. For use in medical treatments, nearly all drugs need to be formulated in order to maximize its therapeutic effect while minimizing the possible side effects. Often times, the therapeutic agent is considered as the functional cargo to be delivered, and its potential as molecular building blocks has been largely ignored. In this work, I detail the use of drugs as effective molecular building units to create well-defined supramolecular nanostructures that contain a 100% drug loading for systemic delivery as well as supramolecular hydrogels for local treatment of human diseases. Our studies show that molecular design, assembly conditions and kinetic pathways are all critical factors that govern the resultant nanostructures and consequently the therapeutic outcomes. These findings have led us to believe that self-assembly of drugs offers an innovative yet simple way for drug formulation, producing drug-based nanomaterials for use in a wide variety of biomedical settings.

562727 Nanotechnology-Based Probes of Structure-Function in Living Tissue

The human brain is highly connected over a number of different length and time scales, spanning communication between neurons to communication across distant brain regions. The study of how the structure of the brain is connected to its many functions is an emerging approach to understanding neurological and psychiatric disorders. However, while advances in brain imaging methods are producing maps of structure or function with unprecedented accuracy and resolution, the ways in which structure shapes function is still not known. Our lab develops methods that can integrate multiple types of data to better understand and predict structural and functional relationships in the brain. In this talk, we will discuss our use of nanotechnology, neurobiology, and data science tools to characterize changes in extracellular matrix (ECM) structure in living brain tissue. We first establish the utility and sensitivity of our toolbox.

Using multiple particle tracking, we track 100,000's of nanoparticles in organotypic brain slices, extract nanoparticle trajectories, and apply machine learning to predict aspects of nanoparticle behavior in tissue slices. For example, we measure nanoparticle size, surface charge, and protein corona formation in vitro, and based on the diffusion of these nanoparticles, we predict whether a particle will internalize or not into cells in the brain. We validate our findings with FACS and immunofluorescent imaging. We next show modulation of organotypic brain slices to capture disease pathways common in many neurological disorders. We use RT-PCR, ELISAs, and live cell imaging to quantify molecular and functional changes in microglia and neurons in response to these induced pathological conditions. Finally, we show how the integration of these techniques can map structural changes of specific ECM structures that regulate neuronal plasticity and the associated functional changes observed in normal development and in the presence of injury.

This work demonstrates that multiple particle tracking of nanoparticle probes in the brain can be used in combination with molecular, cellular, and functional techniques to extract new insights in region-dependent or pathology-dependent structure-function relationships. The long-term goal of this work is to change the way researchers combine diverse data with advanced imaging by integrating outputs from multiple experimental sources using machine learning and data science tools.

570952 Controlled Functionalization of Carbon Nanomaterials for Multifunctional Applications

Carbon nanomaterials, including carbon nanotubes and graphene have emerged as an attractive candidate for various applications due to its unique structure and properties. Like other materials, however, carbon nanomaterial that possesses desirable bulk properties rarely features the surface characteristics required for certain specific applications. Therefore, surface functionalization is essential, and researchers have devised various covalent and non-covalent chemistries for making graphitic materials with the bulk and surface properties needed for various specific applications. Judicious application of these *site-selective* reactions to carbon nanomaterials has opened up a large variety of multifunctional applications, including energy conversion and storage devices, flexible optoelectronics, chemical/bio-sensors, dry adhesives, to name a few.

In this talk, I will provide some rational concepts for the controlled functionalization of carbon nanomaterials for the multifunctional applications, along with an overview on recent progresses in this emerging research area.

586510 Heating, Curing, and Welding of 3D Printed Carbon Nanotube-Polymer Systems by Locally Induced RF Heating

Additive manufacturing through material extrusion (ME), often termed 3D printing, is a burgeoning method for manufacturing thermoplastic components. However, a key obstacle facing 3D-printed plastic parts in engineering applications is the weak weld between successive filament traces, which often leads to delamination and mechanical failure. This is the chief obstacle to the use of thermoplastic additive manufacturing. We have recently demonstrated a novel concept for welding 3D-printed thermoplastic interfaces using intense localized heating of carbon nanotubes (CNTs) by microwave irradiation. The microwave heating of the CNT-polymer composites are a function of CNT percolation, as shown through in situ infrared imaging and simulation. We apply CNT-loaded coatings to 3D printer filament; after printing, microwave irradiation is shown to improve the weld fracture strength to values comparable to injection-molded samples. Recent results indicate that these fields can actually be mounted on the 3D printer head itself, using a dielectric barrier discharge as a non-contact electrode. These remarkable results open up entirely new design spaces for additive manufacturing and also yield new insight into the coupling between dielectric properties and electromagnetic field responses for nano-filled materials. Further study has shown that low-frequency RF fields can also couple with CNT networks, allowing for a variety of rapid scanning techniques without any need for shielding. We have demonstrated that such techniques can be used to induce rapid localized heating in a range of technologies, including thermoset 3D printing and localized thermoset curing, with applications to the automotive sector.

564007 Award Submission: Directed Evolution of the Optoelectronic Properties of Synthetic Nanomaterials

Directed evolution is a powerful approach for tailoring protein properties toward new or enhanced functions. This technique allows one to alter a protein in a directed manner even in the absence of a known relationship between protein structure and function. Several nanocomplexes, such as optical sensors based on DNA-wrapped single-walled carbon nanotubes (DNA-SWCNTs), similarly lack a known structure-function relationship. For example, DNA-SWCNT complexes exhibit variations in their fluorescence that depend on the sequence of the DNA wrapping. However, little is known about the relationship between the DNA sequence and its effects on the fluorescence properties of SWCNTs. In this study, we use directed evolution as a guided approach to tuning the optoelectronic properties of DNA-SWCNT complexes through DNA mutation. Applying this technique over two rounds of DNA mutagenesis and screening, we evolved a DNA-SWCNT sensor with a 56% enhancement in fluorescence intensity. Furthermore, this sensor shows retained selectivity and sensitivity towards the analyte of interest. Though we demonstrate this approach to improve sensor brightness, directed evolution offers far-reaching possibilities in the field that can also be applied to improve additional sensor properties such as sensitivity and selectivity.

575544 Award Submission: Process Intensification for Small Molecule Synthesis Via Recombinant Peptide Templated Palladium and the Elimination of Intermediate Purification

To illustrate the concept of process intensification for small organic molecule synthesis, a project was undertaken to test the catalytic ability of recombinant peptide-templated palladium nanoparticles (Pd NPs) in an impure environment. First, the soluble fraction of bacterial lysates containing repeat units of a metal binding domain (Pd4) fused to the carrier protein Green Fluorescent Protein was used to direct palladium nanoparticle synthesis without purification. Highly ordered NPs on the order of 2 nm were prepared and then used for cross-coupling reactions commonly used in the pharmaceutical industry (e.g. the synthesis of the anti-cancer drug Tykerb). Data showed that Pd NPs without any purification could catalyze the synthesis of fifteen representative compounds via Suzuki or Stille cross-coupling reactions, with the purification of the drug analogs the only purification step of importance. Thus, by using a combination of green chemistry to both produce the catalyst particles and product we illustrate a “one pot process” that requires the minimum number of overall purification steps from start to final product.

571050 Award Submission: Nanoparticle Carrying T Cells for Targeted Drug Delivery

Bionanotechnology (22B 01)

Award Submission: Nanoparticle Carrying T Cells for Targeted Drug Delivery

Muhammad Raisul Abedin[a], Yongjian Liu[b], Sutapa Barua[a,1]

[a] **Department of Chemical and Biochemical Engineering, Missouri University of Science and Technology,**

Rolla, MO-65409

[b] **Mallinckrodt Institute of Radiology, Washington University School of Medicine, Washington University in St. Louis, St. Louis, MO-63110**

[1] **Corresponding Author: Email:** baruas@mst.edu

Abstract

Objective

The overall goal of this article is to explore a combinatorial approach in exploiting the specificity of antibody conjugated drug nanorods with the direction of polyclonal T cells for breast cancer therapy. We hypothesize that the attachment of Trastuzumab-conjugated paclitaxel drug nanorods to natural T cells will enhance their circulation time and substantially alter their anti-tumoral efficacy.

Introduction

Major limitations of the current nano drug delivery vehicles are inefficient transport of drug to the tumor site, uncontrolled nanoparticle-biomolecule interaction at the extra cellular matrix, and intracellular lysosomal or phagolysosomal degradation of the nanoparticle. Moreover, the majority of the designed nano drug vehicles accumulate in the vicinity of the off target cells or tissues resulting in off target toxicity or might be eliminated by the residing macrophages in the organs like liver and spleens. To address this issue, this project aims to take a combinatorial approach of immunotherapies, and targeted drug NPs for efficient delivery of therapeutic payloads to the breast cancer site, and expand therapeutic benefits. Here, we propose an alternative strategy that makes use of the ability of natural immune T cells to circulate and target breast cancer cells for the effective delivery of synthetic drug NPs. Specifically, it is proposed to anchor Trastuzumab conjugated paclitaxel drug nanorods to the surface of patient derived T cells as a means of avoiding the reticuloendothelial system (RES) clearance. We hypothesize that the attachment of Trastuzumab-conjugated paclitaxel drug nanorods to natural T cells will enhance their circulation time and substantially alter their anti-tumoral efficacy.

Materials and Methods

Paclitaxel nanorods (503.42±210 nm) were prepared using the solvent diffusion method. Trastuzumab (TTZ) (Genentech) was conjugated with the functionalized paclitaxel drug nanorods. The antibody drug conjugate (PTXNR-TTZ) was further conjugated with the DOTA chelator for radiolabeling with positron emitting ^{64}Cu isotope. Cytotoxic T lymphocytes were bound with PTXNR-TTZ in serum free un-supplemented RPMI 1640 medium.

Results

The SEM images confirm the cylindrical or rod-like the shape of the paclitaxel drug particles in the nano dimension of (503.42±210) × (96.87±33)nm. The nanoparticle synthesis efficiency was 65%. Monoclonal antibody trastuzumab was chemically conjugated with the paclitaxel drug nanorods with the conjugation efficiency of 53%. The drug NRs along with the antibody conjugated PTXNR-TTZ showed enhanced stability in the biological media. PTXNR-TTZ was bound on the surface of the cytotoxic T lymphocytes without inducing apoptosis to the T cells. T- cell surface bound antibody conjugated drug nanorods induced apoptosis in HER 2+ breast cancer cells resulting in significant tumor clearence.

577828 Award Submission: Systemic Dendrimer-Mediated Delivery of Triptolide Specifically to Tumor-Associated Macrophages Improves Efficacy in Glioblastoma

Introduction: Glioblastoma is among the most severe types of cancers due to challenges with delivering therapies to the brain in addition to traditional solid tumor barriers. The blood brain barrier acts as a natural defense mechanism preventing foreign substances, including medicines, from reaching the brain. In addition, ill-developed vasculature and high interstitial pressure within tumors hampers effective drug penetration. This two-pronged barrier requires the administration of high doses of systemic therapies to achieve clinically relevant quantities in the brain tumor, resulting in significant off-target toxicities. Triptolide is a promising anti-cancer therapeutic that acts both as an immune suppressant in tumor-associated macrophages (TAMs) and as an anti-proliferative in cancer cells. However, like many potent anti-cancer drugs, its off-target activity causes significant systemic toxicities. Therefore, a platform that carries triptolide to the brain and into the brain tumor for triggered release while remaining inactive in the body provides an optimal clinical strategy.

Methods: Our group has shown that hydroxyl-terminated polyamidoamine (PAMAM) dendrimers selectively target activated microglia and macrophages in several neurodegenerative disorders, including TAMs in brain cancers. Here, we present novel PAMAM dendrimers surface modified with triptolide via ester linkages for intracellular and intratumoral triggered release (D-Trip). *In vitro* studies were performed in GL261 murine microglia to assess anti-proliferative activity and in BV2 murine microglia stimulated with IL-4 to explore immune-suppressant activity. *In vivo* experiments were performed in the GL261 mouse model of glioblastoma. Tumor-bearing mice were systemically administered 0.5 mg/kg triptolide or D-Trip daily starting day 5 post-inoculation and monitored regularly for markers of disease progression.

Results: D-Trip released triptolide in a sustained manner over two weeks under intracellular conditions while exhibiting minimal release in extracellular conditions. *In vitro*, D-Trip showed improved immune suppression in TAMs-like activated microglia over triptolide while both performed similarly in inducing tumor cell death. With systemic administration in mice, D-Trip treatment significantly reduced tumor burden ($p = 0.04$ vs. triptolide, $p = 0.02$ vs. control), leading to improved behavioral markers of disease ($p = 0.0012$ vs. triptolide in kyphosis score). D-Trip treatment also ameliorated the significant toxicity at the injection site observed in triptolide treated animals attributed to the targeting and triggered release capabilities of the dendrimer platform.

Conclusions: Here, we present a novel dendrimer-mediated triptolide targeted delivery platform that provides multiple anti-cancer functionalities for improved therapeutic efficacy. Dendrimers specifically target TAMs and release triptolide intracellularly to suppress their pro-tumor polarization. These TAMs then act as depots to release dendrimers into the surrounding tumor microenvironment to induce tumor cell death. These studies demonstrate the significant potential of PAMAM dendrimers for overcoming challenges to brain tumor drug delivery.

562985 Award Submission: Carbon Nanomaterials Enable Plant Genome Engineering without Transgene Integration

Carbon nanomaterials enable plant genome engineering without transgene integration

Food security is threatened by decreasing crop yields and increased consumption in the light of climate change, population growth, and a shortage of arable land. To mitigate these factors, genetic engineering of plants can be employed to create crops that have higher yields and nutritional value, and are resistant to herbicides, insects, diseases, and abiotic stresses such as drought. Despite the recent significant advancements in genome editing (such as CRISPR), most plant species still remain difficult to genetically engineer. The two bottlenecks of generating transgenic plants are (i) efficient biomolecule delivery into plant cells through the rigid cell wall and (ii) the regeneration of transformed tissues[1]. The workhorse method of plant DNA delivery, *Agrobacterium*, limits the range of plant species that can be transformed and results in uncontrolled transgene integration, hence eliciting a genetically modified organism (GMO) regulatory purview of edited plants. To-date, there has yet to be a plant transformation method that enables high-efficiency plasmid DNA delivery, without transgene integration, in a plant species-independent manner for intact plants.

Here, we describe the generation, validation, and optimization of a nanomaterial-based gene delivery platform that can efficiently deliver genes into both model and agriculturally relevant crop plants, without mechanical aid, in a non-toxic and non-integrating manner; a combination of features that is not attainable with existing plant transformation approaches. More specifically, we chemically modified the surface of single-walled carbon nanotubes (SWNTs) with a cationic polymer (polyethylenimine, PEI) to adsorb the negatively charged genetic cargoes via electrostatic attractions. Delivery of plasmid DNA to tobacco, arugula, cotton, and wheat leaves with these modified SWNTs results in strong transient expression of reporter and functional proteins[2,3]. We verified that the transgene does not integrate into plant nuclear genome with a highly sensitive digital droplet PCR analysis[2]. Next, we show chemically modified SWNTs can deliver CRISPR plasmids encoding the nuclease protein Cas9 and guide RNAs targeting genes of interest. Through transient expression of Cas9 and guide RNA in plant cells, we obtain stable editing of endogenous plant genes with efficiencies comparable to *Agrobacterium*-mediated delivery in tobacco leaves. Gene editing rates are quantified using a combination of techniques, such as the restriction site loss assay, Sanger sequencing and TIDE/ICE analysis, and deep amplicon sequencing. We also demonstrate the value of our nano-scale gene delivery platform by editing genes in plant seeds with CRISPR-SWNT delivery, which could eliminate the need for laborious tissue culture protocols to regenerate edited plants. Lastly, we demonstrate the compatibility of nanomaterials for plant biomolecule delivery via RNA sequencing analysis.

Plant genome editing using carbon nanotubes loaded with CRISPR vectors is a breakthrough advancement addressing two crucial bottlenecks in the generation of transgenic plants: (i) gene delivery to enable transient expression without gene integration and (ii) production of stable gene editing without tissue regeneration. Additionally, non-integrating DNA delivery nanotechnologies could enable genetic engineering of plants in a manner that circumvents GMO labeling in the U.S. to facilitate introduction of engineered crops to the market in a time- and cost-effective manner.

1. Cunningham, F.J., Goh, N., Demirer, G.S. et al. Nanoparticle-Mediated Delivery Towards Advancing Plant Genetic Engineering. Cell Press Trends in Biotechnology (2018).

2. Demirer, G.S., Zhang, H. et al. High Aspect Ratio Nanomaterials Enable Delivery of Functional Genetic Material Without Transgenic DNA Integration in Mature Plants. Nature Nanotechnology (2019).

3. Landry M. P., Demirer, G.S. Mature plant transformation with nanoparticle-grafted gene vectors. U.S. Provisional patent filed via UC Berkeley, March 2017.

565576 Award Submission: Self-Assembly of Hydrophobically-Patterned Charged Polypeptides for Drug Delivery

Polyelectrolyte complexation is the result of the interaction of oppositely charged macromolecules in solution. Under precise conditions, these complexes can form liquid-liquid phase separated states called complex coacervates. The complex coacervate phase has a very low interfacial tension with water and therefore can be used as an encapsulant in the food and pharmaceutical industries. Here we design hydrophobically-patterned ionic polypeptides in order to systematically study the effect of hydrophobicity on complex behavior and to assess the ability of these hydrophilic materials to encapsulate hydrophobic drugs. Two generations of oppositely charged peptides are designed, containing non-charged amino acids of increasing hydrophobicity: glycine, alanine, and leucine; while lysine and glutamic acid are used as the charged amino acids for the polycation and polyanion sequences, respectively. Previous work with ionic polypeptides has determined that chirality is an important factor that controls the phase behavior of polypeptide complexes, where alternating patterns of L and D amino acids leads to complex coacervate formation, while homochiral sequences form solid precipitates. Therefore, here we use an alternating sequence of D and L-chiral amino acids in order to promote liquid coacervate formation. The first generation designs consist of D-charged residues and L-hydrophobic residues, while the second generation peptides, also have alternating chiral patterns, but a higher charge density. Peptide sequences are characterized using mass spectroscopy (MALDI-TOF), H NMR and circular dichroism (CD). The polypeptide based polyelectrolyte complexes are characterized using turbidity measurements to evaluate the extent of complex formation as well as the stability of the polyelectrolyte complexes against salt (NaCl) and temperature. In addition, optical microscopy and infrared spectroscopy are used to characterize complex phase behavior and secondary structure. Our results indicate that increasing hydrophobicity improves the stability of the complexes and increases the amount of complex formed with increasing temperature. Moreover, the density and amount of the complex phase are found to increase with hydrophobicity. We use a hydrophobic model dye to assess the encapsulation behavior of the complexes showing that encapsulation efficiency of the complexes increases as the hydrophobicity increases. This work provides insight for the potential application of these complexes in drug delivery containing both charged biological therapeutics and hydrophobic drugs.

569027 Award Submission: Rational Selection of Intestinal Permeation Enhancers for Oral Macromolecular Drug Delivery

Oral delivery of proteins remains a pressing, unmet need in the field of drug delivery, primarily due to poor transport across the intestinal epithelium. A common approach to achieve transepithelial transport is the use of materials called permeation enhancers that manipulate the epithelial cell monolayer. Unfortunately, the clinical translation of permeation enhancers has been limited. Although many reported enhancers increase drug permeability across commonly-used Caco-2 cell monolayers, their efficacy rarely extends to studies in mice or rats. Poor translation from in vitro to in vivo systems can be partially attributed to inadequate understanding of the physical limits of intestinal permeation enhancement in cell culture and in animals. This study systematically determined how three mechanistically distinct permeation enhancers affect the in vitro and in vivo permeability of macromolecules with varied size, surface charge, and conformation. This newfound knowledge will enable rational selection of permeation enhancers when developing delivery systems for therapeutic proteins.

Five proteins (insulin, 5.7 kDa; cytochrome C, 12 kDa; ovalbumin, 44 kDa; bovine serum albumin, 66 kDa; and mouse IgG, 150 kDa) and five dextrans with corresponding molecular weights were fluorescently labeled and characterized for hydrodynamic diameter, molecular weight, and surface charge. Permeability was measured in vitro using Caco-2 monolayers and in mice by intestinal injection. Solutions of one marker mixed with one of three enhancers (1-phenylpiperazine, silica nanoparticles, or sodium deoxycholate) were administered to cells or mice, and fluorescence measurements quantified permeation from media or blood samples.

The three permeation enhancers operated via different mechanisms in Caco-2 monolayers, and the effect of macromolecular size on permeation varied significantly with enhancer type. In Caco-2 monolayers, silica nanoparticles were efficacious for markers less than 40 kDa, while low concentrations of 1-phenylpiperazine increased permeability of all markers. Translation to mice was challenging in that effective in vitro enhancer concentrations were not reflective of required concentrations in vivo. However, with optimization of enhancer doses, significant permeability increases were observed in mice. For 4 kDa dextran, silica nanoparticles more than doubled the permeability and 1-phenylpiperazine achieved a remarkable 17-fold increase in permeability.

This study improves our understanding of how permeation enhancers aid intestinal absorption of macromolecular drugs by thoroughly examining the performance of three mechanistically distinct permeation enhancers. Ultimately, compatibility between enhancer mechanism and the properties of the macromolecules dictated permeation enhancer performance in mice.

569017 High-Yield Scalable Graphene Nanosheet Production from Compressed Graphite Using Electrochemical Exfoliation

Electrochemical exfoliation is a promising bulk method for producing graphene from graphite; in this method, an applied voltage drives ionic species to intercalate into graphite where they form gaseous species that expand and exfoliate individual graphene sheets. However, a number of obstacles have prevented this approach from becoming a feasible production route; the disintegration of the graphite electrode as the method progresses is the chief difficulty. Here we show that if graphite powders are contained and compressed within a permeable and expandable containment system, the graphite powders can be continuously intercalated, expanded, and exfoliated to produce graphene. Our data indicate both high yield (65%) and extraordinarily large lateral size (>30 μm) in the as-produced graphene. We also show that this process is scalable and that graphene yield efficiency depends solely on reactor geometry, graphite compression, and electrolyte transport.

575007 Temperature and Gate Dependent Photoresponse in MoS_2-WS_2 Heterostructures

Van der Waals heterostructures enable the combination of functionalities from different nanomaterials to achieve advanced, superimposed properties. Here, mixed MoS_2-WS_2 heterostructure grains were grown using atmospheric pressure chemical vapor deposition technique, and their optoelectronic properties were investigated. The field effect transistors fabricated from the heterostructure grains exhibited high photoresponsivity, which was influenced by the gating potential as well as the measurement temperature. During the gating potential sweep from -5 V to +55 V, the light to dark source drain current ratio (I_{light}/I_{dark}) decreased from 16 to 3 and from 7 to 2 at 60 K and 120 K, respectively. Such phenomena can be leveraged to realize novel optoelectronic and photovoltaic applications.

571652 Fabrication of Multi-Dimensional Functional Graphene/Mxens and Their Industrial Applications

Owning 2-dimentional conjugated structures, Graphene/MXenes showed many unique physical and chemical properties, which could be further converted to multi-dimensional materials as the 0-dimentional quantum dots, 1-dimensional nanoscrolls/fibers, 2-dimentional coating/papers, 3-dimentional gels/architectures through chemical as well as physical treatments. The conversions brought out novel optical/electrical/magnetical properties which showed great industrial applications potentials. In this presentation, a brief introduction for our research will be given on fabrication and application Multi-Dimensional Functional Graphene/MXenes on energy storage/conversion materials as metal-free ORR electrocatalyst, thermoelectric conversion materials, supercapacitors, optoelectrical materials as transparent conductive films, photochromic/electrochromic materials, environment protection materials as desulfurization and denitrification materials, seawater desalination materials, carbon dioxide capture materials as well as engineering plastics reinforcing agents.

576529 How Graphene Nanoribbons Grow during Chemical Vapor Deposition on Ge(001)

Chemical vapor deposition (CVD) of carbon containing precursors is a convenient way to synthesize graphene on different substrates. Interestingly Ge(001) is one of the few substrates, where – at slow growth rates - elongated graphene shapes with smooth armchair edges, so called graphene nanoribbons (GNRs), can be synthesized [1,2]. Due to their finite bandgap at width below 10 nm, the controlled synthesis of GNRs is highly desirable, but the underlying reasons for the favorable growth behavior on Ge(001) are poorly understood.

In this contribution we study the growth of graphene on Ge(001) during chemical vapor deposition using an empirical kinetic Monte Carlo approach. We find that two different growth regimes exist, which are determined by the competition between the formation of new rows and their growth. Our results indicate that the smoothness of the edges is influenced by row growth rate and the mobility of the graphene precursor species being attached to the graphene edges. Lastly, we find that anisotropy in the formation of this precursor species at different edges, its thermodynamic stability, and the formation of new graphene rows can lead to high aspect-ratio graphene shapes. Comparing our results to experimentally observed growth properties of graphene on Ge(001), we identify the anisotropic thermodynamic stabilization of graphene precursors as the most likely reason for the formation of narrow graphene nanoribbons on this surface. We also demonstrate that kinetic effects impose limits on the obtained aspect ratios.

[1] R.M. Jacobberger, B. Kiraly, M. Fortin-Deschenes, P.L. Levesque, K.M. McElhunny, G.J. Brady, R. Rojas Delgado, S.S. Roy, A. Mannix, M.G. Lagally, P.G. Evans, P. Dejardins, R. Martel, M.C. Hersam, N.P. Guisinger, Nat. Commun. 6, 8006, 2015

[2] R.M. Jacobberger,E.A. Murray, M. Fortin-Deschenes, F. Göltl, W.A. Behn, Z.J. Krebs, P.L. Levesque, F.E. Savage, C. Smoot, M.G. Lagally, P. Desjardins, R. Martel, B. Brar, O. Moutanabbir. M. Mavrikakis, M.S. Arnold, Nanoscale 11, 4864, 2019

566637 Increased Electron Transport inside Microbial Fuel Cell through Interfacing Graphene with *Geobacter*

With the increased interest in developing cost-effective and readily available energy sources, the research in the microbial fuel cell (MFC) field has gained an increased attention. Here, we report the use of graphene's high electron mobility coupled with *Geobacter sulfurreducens'* biocatalyst nature to increase the current response of a MFC. It is well-accepted that one of the limiting factors of current generation of a *Geobacter* based MFC is the electron transport process in the anodic chamber. We attempt to address this limitation by utilizing the enhanced electrical conductivity of graphene. To study the interface of graphene-*Geobacter*, Raman spectroscopy and scanning electron microscopy (SEM) were employed. The Raman characterization studies of the interface revealed an n-doping effect on graphene, with an approximate shift of 5 cm^{-1}. These results demonstrate that the graphene- *Geobacter* interface provides an approach to overcoming the challenges of conventional MFCs.

564760 Wetting Transparency of Graphene on Hydrogel

The water wettability of a solid surface is determined by the solid material and, therefore, is usually changed by either using a different solid material or applying a coating. However, recent research has suggested that water can (at least partially) "see through" an atomic-thick graphene layer, giving rise to the so-called "wetting transparency" effect. This interesting finding provides tremendous opportunity since it means that the wettability of an atomic-thick film can be tuned by selecting an appropriate supporting substrate. The picture becomes even more appealing if the substrate is a liquid instead of a solid, since one can readily change the liquid underneath the graphene, thereby providing real-time control of the surface wettability, a capability highly desirable in many applications. Nevertheless, research to date has almost exclusively focused on the solid substrates and very little is known about the wettability of graphene on liquid substrates. One of the key reasons for that is macroscopic water droplets rupture the graphene supported by a liquid such as water.

Here we report our experimental results on the wetting transparency of graphene supported by a hydrogel. The key advantage of using a hydrogel support is that it contains mainly (e.g., 96%) water yet is mechanically robust enough to support the graphene in the contact angle measurements. The single-layer graphene on Cu grown by chemical vapor deposition (CVD) has been etched and then transferred onto the agarose hydrogel. The Raman spectrum of graphene indicates the high quality of graphene on the hydrogel. The contact angles (CA) of two commonly used testing liquids, water and diiodomethane, on both bare hydrogel and graphene/hydrogel suggest that graphene is almost complete wetting transparent, i.e., the CA of the hydrogel is very close to that of graphene/hydrogel. The possible mechanisms for the observed wetting transparency will also be discussed.

570891 Metal-Free Carbon Electrocatalysts for Efficient Energy Conversion and Storage

Green and renewable energy technologies, such as fuel cells, batteries, and water-splitting systems, hold great promise to solve current energy and environmental challenges. However, noble metal catalysts (*e.g.*, Pt, Pd, RuO_2, IrO_2) are generally needed to promote the *hydrogen evolution reaction* (HER) for hydrogen fuel generation from photo-electrochemical water-splitting, *oxygen reduction reaction* (ORR) in fuel cells for energy conversion, and *oxygen evolution reaction* (OER) in metal-air batteries for energy storage. The high cost of precious metal-based catalysts and their limited reserve have precluded these renewable energy technologies from large-scale applications.

Along with the recent intensive research efforts in non-noble metal based ORR catalysts, we have previously demonstrated that vertically-aligned nitrogen-doped carbon nanotubes could actively catalyze ORR *via* a four-electron process free from the CO poisoning effects with a 3-time higher electrocatalytic activity and better long-term durability than that of commercial Pt/C catalysts. The improved catalytic performance was attributed to the doping-induced charge transfer from carbon atoms adjacent to the nitrogen atoms to change the chemisorption mode of O_2 and to readily attract electrons from the anode for facilitating the ORR.

Subsequently, it was demonstrated that various graphitic carbon materials, doped with heteroatoms of different electronegativities from that of carbon atom, physically adsorbed with certain polyelectrolytes, and even without any apparent dopant or physically adsorbed polyelectrolyte, could also exhibit good ORR performance. More recent studies have further demonstrated that certain heteroatom-doped carbon nanomaterials could act as metal-free bifunctional catalysts for ORR/OER in metal-air batteries for energy storage, and even ORR/OER/HER trifunctional catalysts for self-powered water-splitting to generate hydrogen fuel and oxygen gas from water.

In this talk, I will summarize some of our work on the metal-free catalysts based on carbon nanomaterials for various energy-related reactions, along with an overview on the recent advances and perspectives in this exciting field.

570849 Nanoparticle Applications for Catalysis

The distinct optical, electronic and catalytic properties of colloidal gold nanoparticles (AuNPs) which are not seen in bulk gold have made them an attractive research subject in the past few decades. Oxidation of alcohols, reduction of nitroarenes, hydrogenation of quinolone and carbon - carbon coupling reactions are the reactions oxidized by colloidal (AuNPs). The preference for supported catalysts in industries over colloidal catalysts is due to their usability in continuous reactors and simplicity in recovery through simple separation methods like centrifugation, sedimentation and filtration for reuse. In comparison, recovery of colloidal catalysts is a bit complicated, employing methods like temperature or pH induced aggregation and redispersion, solvent extraction etc. However, the prospective advantages of colloidal (AuNPs) over supported catalysts lie in improved catalytic activity, selectivity and milder reaction conditions with applications in chiral catalysis. The colloidal (AuNPs) need to be stabilized with ligands to remain dispersed and be functional as catalysts. This also causes surface passivation of the nanoparticles thereby reducing their catalytic activity. We are exploring the effects of the organo thiol stabilizing ligands on the catalytic activity of colloidal (AuNPs). Reduction of 4-nitrophenol to 4 -aminophenol with sodium borohydride ($NaBH_4$) as a reducing agent catalyzed by colloidal (AuNPs) is our model reaction. Effects of the chain length of ligands and surface coverage of nanoparticles on catalytic activity are studied. Studying the relationship between the molecular structure of the stabilizing agents and the surface coverage of the nanoparticles is vital for understanding the mechanism of the catalytic rate of colloidal (AuNPs) for redox reactions. The use of pH sensitive ligands for pH triggered recovery and reusability of colloidal (AuNPs) catalysts is also studied.

574610 Investigation of Gel Polymer Electrolyte Failure Mechanism in Lithium-Sulfur Batteries

Investigation of gel polymer electrolyte failure mechanism in Lithium-Sulfur batteries

Somayeh Zamani, George Shebert and Yong Lak Joo

2019 AIChE Annual Meeting

Lithium sulfur (Li-S) batteries are a promising candidate for next-generation energy storage due to their high theoretical energy density. However, well-known challenges such as polysulfide crossover, large volume changes during cycling, lithium dendrite growth, and passivation of reaction sites limit the capacity and capacity retention of Li-S batteries. One approach to address these issues is to use gel polymer electrolyte (GPE) in place of the typical liquid electrolyte. It is believed that GPE can reduce the dissolution and diffusion of the intermediate polysulfides and enhance the capacity retention in Li-S batteries, but additional research is needed. In particular, the difference in failure mechanisms between GPE and liquid electrolyte cells is not well understood, which is important knowledge for the rational design of high performance GPE cells.

We have investigated the effects of gel polymer electrolyte on lithium-sulfur coin cell failure mechanisms. In particular, we used a discharge test with pauses to probe how the choice of electrolyte affected the amount of capacity recovered due to the pause. After discharging normally until failure, the cell was paused for some time, after which discharge was restarted until the second and final failure. Gel polymer electrolyte was found to have a higher percentage capacity recovery after pausing than liquid electrolyte, and that recovery improved at higher C-rates. Large percentage capacities recoveries suggest that mass transport limitations are leading to cell failure rather than cathode passivation as is frequently proposed. To help explain this behavior, we use a 1-dimension + time numerical continuum model for the Li-S full cell that includes the pausing step. The model includes the electrochemical reactions and precipitation of sulfur and intermediate polysulfide species, transport of species through the battery, and the effect of mass transport limitations on cell failure. The simulations show that variations in the bulk polysulfide diffusion coefficients can explain many of the differences between GPE and liquid electrolyte experimental results, and further confirm that mass transport limitations are a critical obstacle in improving the performance of GPE Li-S batteries.

562069 Design and Engineering of Interfaces/Interphases for Stable Li Anodes

Li metal and Li-alloy based anode materials are the most promising anodes for next-generation Li batteries. The poor interfacial stability (unstable solid-electrolyte interphase (SEI)) in the battery has been the primary issue hindering their practical application. In this talk, I will present a strategy to reinforce the SEI with desired properties including good tolerance to the Li-based material volume change and efficient surface passivation against electrolyte penetration. The strategy works via introducing multiple functional components bonded to the Li-based material surface into the SEI. The SEI reinforced shows much better stability than the SEI reinforced by electrolyte additive strategy, which is the current state-of-art and commercially used solution to SEI stability issue.

577527 Mechanistic Elucidation and Electrochemical Performance of Amorphous Carbon Interfaced Graphite-Silicon Advanced Anodes

Graphite, inspite of being the mainstay in the lithium-ion battery (LIBs) industry for decades, has come to realize its own limitation of theoretical anode capacity (372 mAh g^{-1}).[1] Due to this reason, there has been a large momentum towards fabrication or invention of new anode materials with improved specific capacity and stable cycle performance. Alloys are a possible answer to the situation and have received much attention. Sn, Ge or Si alloys have higher specific capacity as their atomic framework aid the reaction as compared to the intercalation materials. Amongst all alloys types, silicon has garnered interest as an anode material for the next generation LIBs, owing to its high capacity, however achieving long stable cycle life is a significant challenge due to fatal and critical technical issues (e.g. volume expansion, excess solid electrolyte interface generation, particle fragmentation and pulverization, etc.) and hence impeding the practical use of Si anodes.[2] For commercialization, various approaches have been tried to overcome the issues like fabricating nanostructured silicon, coating or dispersing Si-nanoparticles (Si-NPs) with active or less active materials for stress reduction, generated due to volume expansion. Though these materials with better electrochemistry are quite promising, there are still several problems that need to be addressed like low tap density, uneconomical and complex processing, accelerated side reactions due to the large surface area and poor scalability.[3]

To address these challenges, we engineered graphite-silicon composite bridged using wheat flour-derived amorphous carbon (GCSi) for realizing high performance LIBs.[4] GCSi composite utilizes the carbon from the wheat source to effectively confine the volume expansion and improve the electrical conductivity by developing an interface between the graphite and silicon electrode. The synthesis process uses an easily scalable pyrolysis technique in an inert atmosphere to homogenize Si-NPs, wheat flour and graphite via ball-milling. Cross sectional scanning electron micrographs of lithiated and delithiated electrodes revealed the volume changes. Compared to graphite-Si control sample, GCSi had lower volume changes suggesting accommodation of the volume expansion of Si-NPs. STEM and HAADF studies confirmed that the wheat flour derived carbon in the GCSi composite physically bridges the graphite and Si NPs. Mechanistic elucidation of the tailored composite was achieved with *in-situ* HR-ETEM as a function of temperature. The mixture was deposited on a MEMS ETEM heating chip, and an area was selected where graphite, wheat flour, and silicon overlapped, with the Si-NPs buried in a cluster of flour. The cluster sat on top of a sheet of graphite - all suspended over vacuum for high resolution imaging. Temperatures were ramped up from 40°C to 500°C at 12 °C min^{-1} with images captured every 5 minutes. This ramp and imaging allow the capture of the material transition *in-situ*. The carbonization and dehydrogenation of the starch was observed at 400 °C and completed at 500 °C. As the starch homogenized, it appeared to "fuse" into the graphite. The inner defects could be seen to transition into a uniform amorphous carbon mass because of conversion of carbohydrate mass to carbonaceous entity. The BF images show the cross section of the sample as the process occurred, showing the inner structure homogenizing. The Secondary Electron recording showed as the surface of the starch goes from highly course to a smooth continuous surface. EDX was used to confirm the elemental composition of the particles during the transition. This network generated by the wheat carbon, encapsulating the Si-NPs, lead to the formation of a conductive and physical cage that results in an improved and stable performance of the Si-based anodes.

The designed GCSi composite architecture comprising about 25 wt% silicon showed a high initial discharge capacity of >1000 mAh g^{-1} with high coulombic efficiency. GCSi composite anode also demonstrated exceptional rate capabilities. This stability and high performance are a result of the improved mechanical stability and improved electrochemical kinetics made possible by well interconnected composite architecture of the wheat derived amorphous carbon. Thus, the proposed strategy of incorporating an amorphous carbon network into the G-Si composites decreases the inherent issues associated with the pristine Si electrodes and provide an alternative solution for the development of advance Si-based anodes. The mechanistic elucidation of interfacial development of amorphous carbon bridging standard graphite and silicon as a function of temperature yielding advanced anode architecture will be discussed in great details using HRTEM and *in-situ* videos.

References:

(1) Goriparti, S.; Miele, E.; De Angelis, F.; Di Fabrizio, E.; Proietti Zaccaria, R.; Capiglia, C. Review on Recent Progress of Nanostructured Anode Materials for Li-Ion Batteries. *J. Power Sources* **2014**, *257*, 421–443.

(2) McDowell, M. T.; Lee, S. W.; Nix, W. D.; Cui, Y. 25th Anniversary Article: Understanding the Lithiation of Silicon and Other Alloying Anodes for Lithium-Ion Batteries. *Adv. Mater.* **2013**, *25* (36), 4966–4985.

(3) Tang, J.; Dysart, A. D.; Kim, D. H.; Saraswat, R.; Shaver, G. M.; Pol, V. G. Fabrication of Carbon/Silicon Composite as Lithium-Ion Anode with Enhanced Cycling Stability. *Electrochim. Acta* **2017**, *247*, 626–633.

(4) Parekh, M. H.; Parikh, V. P.; Kim, P. J.; Misra, S.; Qi, Z.; Wang, H.; Pol, V. G. Encapsulation and Networking of Silicon Nanoparticles Using Amorphous Carbon and Graphite for High Performance Li-Ion Batteries. *Carbon N. Y.* **2019**, *148*, 36–43.

558587 Silicon Carbide As a Protective Layer to Stabilize Si-Based Anode By Inhibiting the Chemical Reaction

The poor cycle performance of the silicon based (Si-based) anode hindered its practical application. Here, a fresh and fundamental mechanism was proposed to interpret the rapid capacity decay of Si-based materials. Silicon could chemically react with $LiPF_6$ and generate Li_2SiF_6 aggregations constantly during cycling process. What's worse, nano carbon coated on the silicon acted as a catalyst to accelerate such detrimental reaction. By taking advantages of high strength and toughness of silicon carbide (SiC), SiC layer was rationally introduced between inner silicon and outer carbon to inhibit the chemical formation reaction of Li_2SiF_6. The side reaction rate reduced significantly attributed to the increased activation energy of such reaction. The Si@SiC@C maintained a specific capacity of 980 mAh g^{-1} at the current density of 1 A g^{-1} after 800 cycles with the initial Coulombic efficiency over 88.5%.

570524 Free-Standing Bimetallic Sulfide@Carbon Nanofiber Networks As Lithium-Ion Battery Anodes

Anode materials play a crucial role in the overall electrochemical performance of lithium-ion batteries (LIBs). In this work we use electrospinning, a straight-forward, scalable, and low-cost technique, to embed nickel cobalt sulfide nanoparticles into carbon nanofiber networks. After obtaining fresh nanofibers via electrospinning, the fresh fiber mat is simultaneously carbonized and sulfurized under the nitrogen atmosphere at 700°C for 1 hour, where we obtained the free-standing nonwoven carbon nanofiber networks embedded with metal sulfide nanoparticles subsequently. The as-prepared materials are highly stable with numerous pores among the fiber networks, allowing for fast electrolyte penetration. When tested as the anode, the materials exhibit high reversible capacity, superior capability, and excellent rate performance. These excellent performances can be attributed to the fact that the nonwoven 3D fiber networks, not only do they provide more active sites but also are highly resistant to volume expansion. Additionally, the synergistic effect of two metal sulfides enhances the performance of lithium-ion storage. Furthermore, this free-standing and binder-free electrode also avoids tedious slurry making steps and eliminates uncontrolled structure formation, low porosity, and undesirable interfaces caused by involving binder and the extra substrate. Ultimately, we reveal a facile method to design LIBs anode materials with excellent performance.

575949 Award Submission: Quantum Capacitance Based Amplified Graphene Phononics for Studying Neurodegenerative Diseases

Amyotrophic lateral sclerosis (ALS) is the most common adult-onset motor neuron disease, characterized by a rapid loss of upper and lower motor-neurons resulting in patient death from respiratory failure within 3-5 years of initial symptoms onset. Although at least 30 genes of major effect have been reported, the pathobiology of ALS is not well understood. Compounding this is the lack of a reliable laboratory test which can accurately diagnose this rapidly deteriorating disease. Herein, we report on graphene's phonon vibration-energies as a sensitive measure of the composite dipole moment of the components of the interfaced cerebrospinal fluid (CSF) to specifically identify patients with ALS disease. The second-order overtone of in-plane phonon vibration energy (2D) of graphene shifts by 3.2 ± 0.5 cm^{-1} for all ALS patients studied in this work. Further, the amount of *n*-doping induced shift in phonon energy of graphene, interfaced with CSF, is specific to the investigated neurodegenerative disease (ALS, Multiple Sclerosis and Motor Neuron Disease). By removing a severe roadblock in disease detection, this technology can be applied to study diagnostic biomarkers for researchers developing therapeutics and clinicians initiating treatments for neurodegenerative diseases.

566738 Lipid Membrane Bilayer Destabilization By Amphiphilic Coumarin: A Biophysical Study for Cancer Applications

The interaction between the phospholipid cell membrane and the drug molecules can be quantified by various biophysical and biochemical methods. This study is important is understanding the pharmacology of novel molecules. Though the phospholipid membrane is not the final target of the drug molecule, a knowledge on this initial interaction is necessary to predict the efficacy of the drug. The penetration efficiency of a hydrophilic molecule can be enhanced by inducing amphiphilicity to it.

We have investigated the binding and membrane penetration ability of fluorescent amphiphilic amino coumarin molecules using model cell membranes or liposomes. Firstly, the binding of the molecules to the lipid bilayer was established by surface charge and fluorescence spectroscopy. Then the cell membrane perturbation by these molecules was analysed by examining the phase behaviour of the phospholipids. Based on this, we proposed a combination mechanism of carpet and flip-flop through which these molecules destabilize the membrane. The proposed mechanism is attributed to the induced amphiphilicity in the coumarin molecules and is dependent on concentration. These results conform to the preliminary biological in-vitro activity. These fluorescent amphiphilic molecules have their proposed application in cancer diagnosis and therapeutics.

570219 Award Submission: Engineering Muco-Inert Nanoparticles for Improved Vaginal Drug Delivery during Pregnancy

Introduction: Vaginal drug delivery is the preferred route for targeting the female reproductive tract. By avoiding the harsh gastrointestinal environment and hepatic first-pass effect, local tissue concentration can be increased [1]. Additionally, the uterine-first-pass effect leads to accumulation of vaginally administered drugs in the uterus [2]. Vaginal formulations have been demonstrated to have some efficacy in prevention of preterm birth (PTB), though the optimal formulation for vaginal drug delivery has yet to be explored. Key barriers to effective vaginal drug delivery, including the protective mucus that acts as a barrier and facilitates clearance, must be considered [3]. By identifying physicochemical properties for optimal nanoparticle penetration of CVM, improved vaginal drug absorption can be achieved. Here, we characterized the structural and barrier properties of cervicovaginal mucus (CVM) during pregnancy using fluorescent probe nanoparticles of various sizes and surface coatings. We further utilize these design criteria to formulate drug nanoparticles that can penetrate the vaginal mucus barrier for more effective prevention of PTB.

Methods: 500 CVM samples (spaced monthly) were self-collected by 100 pregnant women. CVM microstructure was characterized by a quantitative multiple particle tracking method. Fluorescently-labeled nanoparticles with and without mucoinert coatings were added to CVM. Particle motions due to thermal energy were recorded and quantified to estimate the mesh spacing of mucin proteins comprising CVM. Further, these nanoparticles were administered vaginally to pregnant mice and the *in vivo* distribution was imaged. Several drugs were formulated into nanosuspensions via wet milling. The size and zeta potential of the formulations were measured. Drug levels the female reproductive tract tissues were analyzed using LC-MS/MS methods. Finally, these formulations were tested in a murine model of PTB. Pregnant mice were challenged with an intra-uterine (IU) lipopolysaccharide (LPS) injection on embryonic day 15 (E15, out of 19 day gestation). Treatment was dosed vaginally daily from E15-18. Mice were checked every

Results: Similar to what was observed previously in CVM from non-pregnant women, nanoparticles with mucoinert coatings have increased mobility in CVM from pregnant women. However, CVM from pregnant women causes a significant reduction in mobility of nanoparticles, particularly for nanoparticles 500nm in size or greater, as pregnancy progresses. Mucoinert nanoparticles ~250nm in size were slowed, but a significant fraction were still highly mobile in CVM. This inferred reduction in pore size in CVM is likely due to increasing progesterone levels throughout pregnancy. Pharmacokinetic studies revealed that nanosuspensions with a size <300 nm and a zeta potential between 0 mV to -5 mV were better able to deliver drugs to the female reproductive tract tissues. Additionally, vaginally delivered drugs were shown to target the female reproductive tract as compared to systemically administered formulations. We were able to formulate Trichostatin A, progesterone, hydroxyprogesterone caproate, vorinostat, and several other drugs into nanosuspensions. Progesterone and its analogs are prescribed for the prevention of PTB clinically. Surprisingly, progestins were only able to prevent PTB in our model when in the presence of a histone deacetylase inhibitor. Analysis of cervical and myometrial tissues revealed anti-inflammatory effects from the combination dosing.

Conclusion: Here, we gained valuable insight for engineering targeted vaginal drug treatments for pregnant women. We have demonstrated the first prevention of preterm birth in this animal model, and are able to bring the dams to deliver live pups. Effectively delivering drugs to target tissues can help to reveal mechanism behind disease. Our technology has allowed us to better target the female reproductive tract and understand mechanisms of preterm birth.

1. Ensign, R. Cone, J. Hanes, JCR 2014, 0:500-5:14
2. De Ziegler, C. Bulletti, B. De Monstier, A.S. Jääskeläinen, Ann N Y Acad Sci, 1997; 828:291-9.
3. Yu, J. Chisholm, W.J. Choi, A. Anonuevo, S. Pulicare, W. Zhong, M. Chen, C. Fridley, S.K. Lai, L.M. Ensign, J.S. Suk, J. Hanes, Adv Healthc Mater, 5 (2016) 2745-2750.

4. C. Condon, P. Jeyasuria, J.M. Faust, J.W. Wilson, C.R. Mendelson, PNAS 2003, 100.16, 9518-9523.

575831 Award Submission: Xeno Nucleic Acid Nanosensors for Enhanced Stability

The omnipresence of salts in biofluids creates a pervasive challenge in designing sensors suitable for *in vivo*applications. Fluctuations in ion concentrations have been shown to affect the sensitivity and selectivity of optical sensors based on single-walled carbon nanotubes wrapped with single-stranded DNA (ssDNA-SWCNTs). We herein observe fluorescence wavelength shifting for ssDNA-SWCNT-based optical sensors in the presence of divalent cations at concentrations above 3.5 mM. In contrast, no shifting was observed for concentrations up to 350 mM for sensors bioengineered with increased rigidity using xeno nucleic acids (XNAs). Transient fluorescence measurements reveal distinct optical transitions for ssDNA- and XNA-based wrappings during ion-induced conformation changes, with XNA-based sensors showing increased permanence in conformational and signal stability. This demonstration introduces synthetic biology as a complementary means for enhancing nanotube optoelectronic behaviour, unlocking previously unexplored possibilities for developing nano-bioengineered sensors with augmented capabilities.

574258 Award Submission: Development of a Model Placental Lipid Bilayer

The placenta has one of the most important roles in controlling fetal development as it is a semi-permeable barrier controlling the exchange of nutrients and wastes. However, it remains one of the least understood organs in the human body. There is a lack of placenta representative *in vitro* and *in vivo* models, making studies of this organ difficult. Meanwhile, during pregnancy a number of molecules may come in contact with the placenta (e.g. environmental toxins, pharmaceuticals, etc.). For example, only approximately 2% of medications approved by the FDA have sufficient data on their risk when used during pregnancy. However, 90% of women take at least one medication throughout the course of pregnancy. Additionally, environmental toxins, such as phthalates are of concern during pregnancy. Previous studies have suggested that women who are exposed to high concentrations of di-(2-ethylhexyl) phthalate (DEHP), were 60% more likely to lose a pregnancy compared to those with lower concentrations. Lipid bilayers have been used to model biological interfaces, such as the liver, red blood cell, and myelin. Here, we sought to develop a placental model. The goal of this work was to develop an *in vitro* placenta model using the placental lipid composition.

First, we investigated the lipid composition of placental (trophoblast) cell lines. We extracted the lipid composition from HTR-8, TCL-1, and primary trophoblast cells using the Bligh-Dyer method. HTR-8 cells are representative of first-trimester trophoblast cells while TCL-1 cells are representative of third-trimester trophoblast cells. After extraction, liquid chromatography with tandem mass spectrometry (LC-MS/MS) was used to quantify the composition of the most abundant lipid classes including phosphatidylcholine (PC), phosphoethanolamine (PE), phosphatidylinositol (PI), phosphatidylserine (PS), and sphingomyelin (SPH). We observed statistically significant differences between the lipid composition of HTR-8, TCL-1, and primary trophoblasts. With the known compositions, we then developed synthetic lipid vesicles using the dry lipid thin film technique and extrusion through a 100 nm polycarbonate membrane with the same lipid composition as the HTR-8, TCL-1, and primary cells. Using dynamic light scattering we confirmed the hydrodynamic diameter of our trophoblast lipid vesicles to be 109 ±1.3 nm. In order to develop a supported lipid bilayer, we used quartz crystal microbalance with dissipation monitoring (QCM-D) to measure frequency (ΔF) and dissipation (ΔD) changes occurring on a silica surface upon introduction of these synthetic vesicles. With this complex vesicle composition, the vesicles adsorb on the surface but do not spontaneously rupture ($\Delta F = -45.4 \pm 13.0$ Hz, $\Delta D = 5.3 \times 10^{-6} \pm 1.6 \times 10^{-6}$). We have synthesized an α-helical peptide that is derived from the N-terminal amphipathic helix of the hepatitis C virus NS5A protein, previously used to rupture complex vesicles into supported lipid bilayers. Using this peptide, we are able to rupture our trophoblast representative vesicles into lipid bilayers on QCM-D ($\Delta F = -30.0 \pm 4.4$ Hz, $\Delta D = 1.3 \times 10^{-6} \pm 0.6 \times 10^{-6}$) and have confirmed the removal of this peptide after rinsing the formed bilayer. Thus, we have successfully fabricated lipid bilayers mimicking the lipid composition of placental studies. Future studies will investigate how environmental toxins (di(2-ethylhexyl) phthalate, mono(2-ethylhexyl) phthalate) and small molecules (folic acid, caffeine, etc.) interact with these bilayers. This work will improve placental understanding so that risk to a developing fetus can be minimized.

580318 Award Submission:Sequence and Aggregation State Determine Stability, Endosomal Processing, and Long-Term Fate of DNA-Wrapped Carbon Nanotubes in Mammalian Cells

Single-walled carbon nanotubes (SWCNTs) have attracted much attention for applications in biomedical imaging and sensing due to their photostable, environmentally responsive near-infrared fluorescence. Single-stranded DNA was found to non-covalently disperse SWCNTs in water while preserving their optical properties and providing a biocompatible wrapping suitable for cellular interactions. The hybridization of DNA-SWCNTs has been studied extensively to determine the relationship between oligonucleotide sequence and SWCNT properties *in situ*, however intracellular environments introduce harsh conditions that can change the identity of the hybrid nanomaterial, altering its intrinsic optical properties. Aggregation via protein-nanotube interactions adds complexity to the biological system, while also causing concerns for cellular toxicity. Here, we show that the stability of DNA-SWCNTs after internalization by mammalian cells is dependent upon the oligonucleotide sequence that wraps the nanotube. The relationship between sequence length, fluorescence stability, and uptake are examined to provide insight on intracellular mechanisms that require considerations in nanotube sensor design. We show that aggregation not only alters the expected rate of uptake into the cells but further affects the endosomal processing pathways. We propose a model for the uptake and expulsion of SWCNTs that accounts for the aggregation state and DNA wrapping. These findings provide fundamental understanding of the interactions between nanotubes and live cells which can be applied towards development of carefully engineered carbon nanotube sensors.

570889 Award Submission: Real-Time Detection of H_2O_2 Signaling in Wild-Type Plants with Nanobionics Approach

As a consequence of their sessile nature, plants have developed complex signaling mechanisms to cope with the highly dynamic environment that they live in. Recent studies revealed that in response to external stimuli, plants can utilize various signaling pathways for cell-to-cell communication to engage different metabolic and genetic machineries involved in the plant defense system. Among the many signaling molecules uncovered, hydrogen peroxide (H_2O_2) represents one of the most versatile signaling molecules in plants in response to environmental stresses such as mechanical wounding. Herein, we develop a nanomaterial-based approach, termed plant nanobionics, using single-walled carbon nanotube (SWNT)-based nanosensors to monitor the wound-induced H_2O_2 signaling pathway in plants. This approach enables non-destructive, real-time detection of changes in endogenous H_2O_2 level in plants at a standoff distance. We demonstrate the versatility of this method to investigate the dynamics between H_2O_2 and electric signal in both model and non-model plants, and show that the nanosensor readout can be paired with portable and inexpensive electronic device. A mathematical model describing the self-propagating mechanism of the wave will be presented. Our work provides a promising tool to elucidate the complex plant signaling pathways across species to aid botany research and precision agricultural applications.

573172 Atomically-Precise Moiré Fringes in 2D Van Der Waals Heterostructures of Graphene and Hexagonal Boron Nitride

The van der Waals force-bound heterostructures of isostructural and isoelectronic graphene and hexagonal boron nitride (h-BN) offer a platform for fundamental phenomena evolving from the resultant Moiré superlattices. Such phenomena in two-dimensional (2D) planar sp^2 lattice stacks of graphene/h-BN or graphene/h-BN/graphene or h-BN/graphene/h-BN are sensitive to their atomic-scale assembly. Current techniques rely on physical or chemical transfer of both h-BN and graphene to build the 2D sandwich heterostructure circuits. These transfer-steps not only introduce the interfacial polymeric and metallic heteroatom impurities but also creates random or mis-aligned regions in the 2D-heterostructures. Here, we report one-dimensional (1D) Moiré fringes in the graphene and h-BN heterostructures developed through direct stacking of graphene and h-BN *via* a bottom-up all-chemical vapor deposition (CVD) strategy. Polycrystalline layer of h-BN is first synthesized *via* a surface chemical-interaction guided process on silicon-based gate-dielectric substrate. Then the graphene layer is grown *via* grain-boundary diffusion of catalytically produced carbon radicals through a polycrystalline metal film on h-BN. Finally, removing the thin metal film (and any graphene on it) results in graphene/h-BN heterostructures with a sharp, defect-free, and atomically-precise interfaces; as confirmed by scanning resonant Raman spectroscopy and X-ray photoelectron spectroscopy studies. The directly-grown, van der Waals bound graphene/h-BN heterostructures exhibit 1D Moiré fringes with a periodic pitch of ˜10 Å at a rotation angle of 16°. Further, the low-temperature transport measurements reveal the charge carrier mobility of 100 $cm^2v^{-1}s^{-1}$, 2-orders of magnitude higher than the graphene/h-BN heterostructure produced *via* co-segregation process. Futuristically, this all-CVD direct-growth strategy will be a transformative approach for scalable production of van der Waals heterostructures to realize complex 2D circuitry and fundamental physics phenomena.

578198 UV Surface Plasmon Resonance Modification By Graphene Pi Plasmon Resonance

UV surface plasmon resonance modification by Graphene Pi plasmon resonance

Xueling Cheng[1], Sourangsu Banerji[1], Jieying Mao[2], Sara Arezoomandan[1], Ting Zhang[2], Steve Blair[1] Berardi Sensale-Rodriguez[1] and Yunshan wang[3]

1. Department of Electrical and Computer Engineering, University of Utah
2. Department of Physics and Astronomy, University of Utah
3. Department of Chemical Engineering, University of Utah

Despite of increasing understandings of UV plasmonic materials, materials that can enable active tuning of UV plasmonic resonance has not been reported. Here, we demonstrate modification of UV SPR on an aluminum (Al) hole-array by coupling Graphene π plasmon resonance with Al SPR. Graphene monolayer exhibits an abnormal absorption peak in the UV region (270-290nm) due to π plasmon resonance. The location and intensity of the absorption peak depends on the position of Fermi-level, which can be adjusted by electric or chemical doping. Al SPR is shown here to be modified by coupling Graphene π plasmon resonance with Al SPR.

FDTD simulation shows modification of Al hole-array transmission by adding a single layer of Graphene on top. The shifts of transmission dips after adding a Graphene layer shows a distinct transition at around the Graphene π plasmon position. For transmission dips that are located at shorter wavelength compared to Graphene π plasmon, up to 8nm blue shifts occur after adding Graphene. On the other hand, up to 20nm red shifts occur for transmission dips that are at longer wavelength relative to Graphene π plasmon(fig.1). This change in the sign of shifts of transmission dips in Figure.2 corresponds to the change in the sign of the real permittivity of Graphene (fig.3). The amount of shifts diminishes as the transmission dip moves further away from Graphene π plasmon resonance into the visible spectrum. Experimentally we have observed redshifts of SPR dips but not blue shifts possibly due to poor light collection below 250nm. Besides, FDTD simulations shows enhanced near-field intensity after coupling Graphene with Al hole array. The observed tunability and the enhanced local electric fields can find applications in tunable UV LED, label free biosensors, etc.

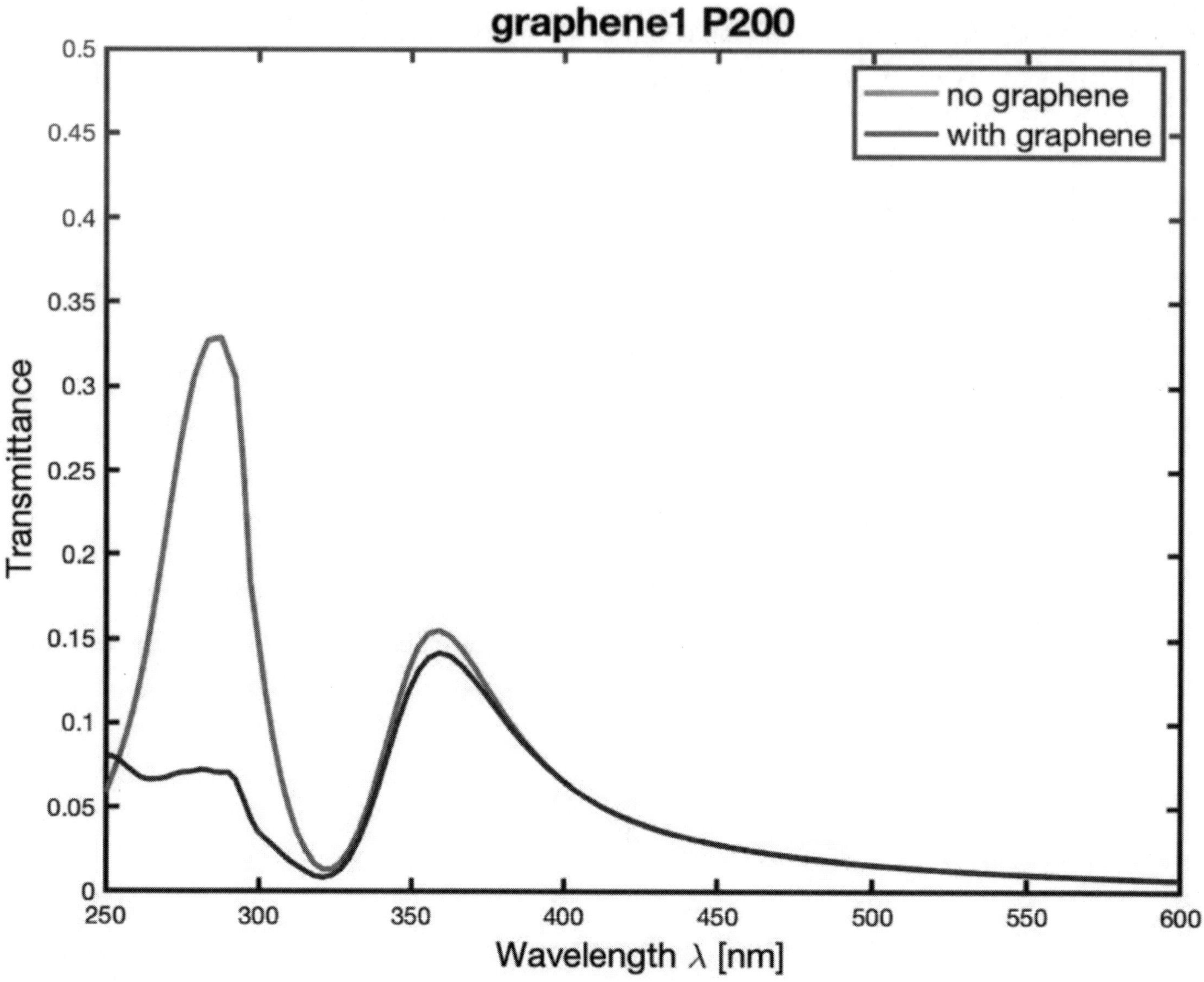
graphene1 P200
no graphene
with graphene
Transmittance
0
0.05
0.1
0.15
0.2
0.25
0.3
0.35
0.4
0.45
0.5
250
300
350
400
450
500
550
600
Wavelength λ [nm]

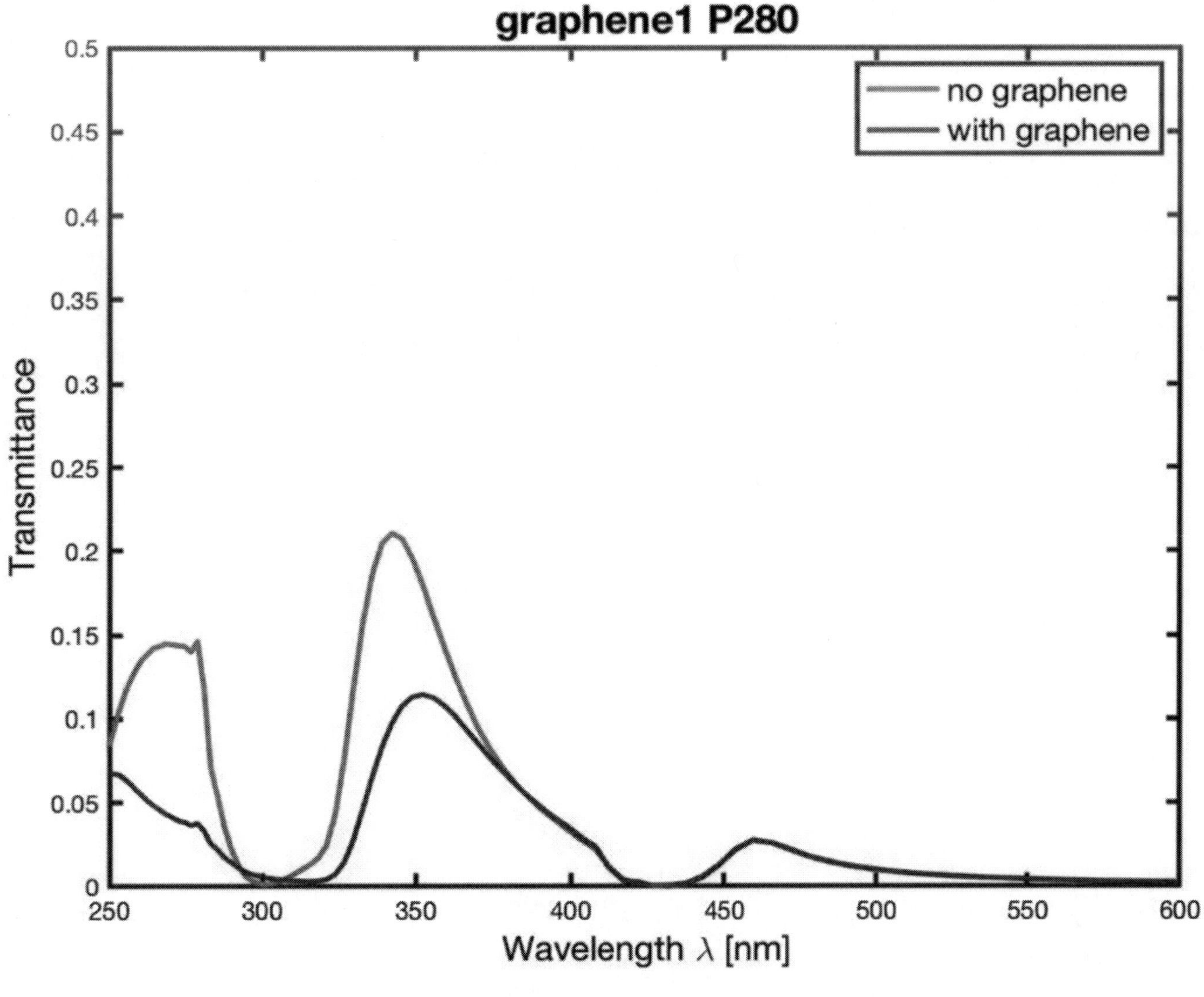

Figure 1. Al hole array with or without graphene for hole array period equals to 200nm (left) and 280 nm (right)

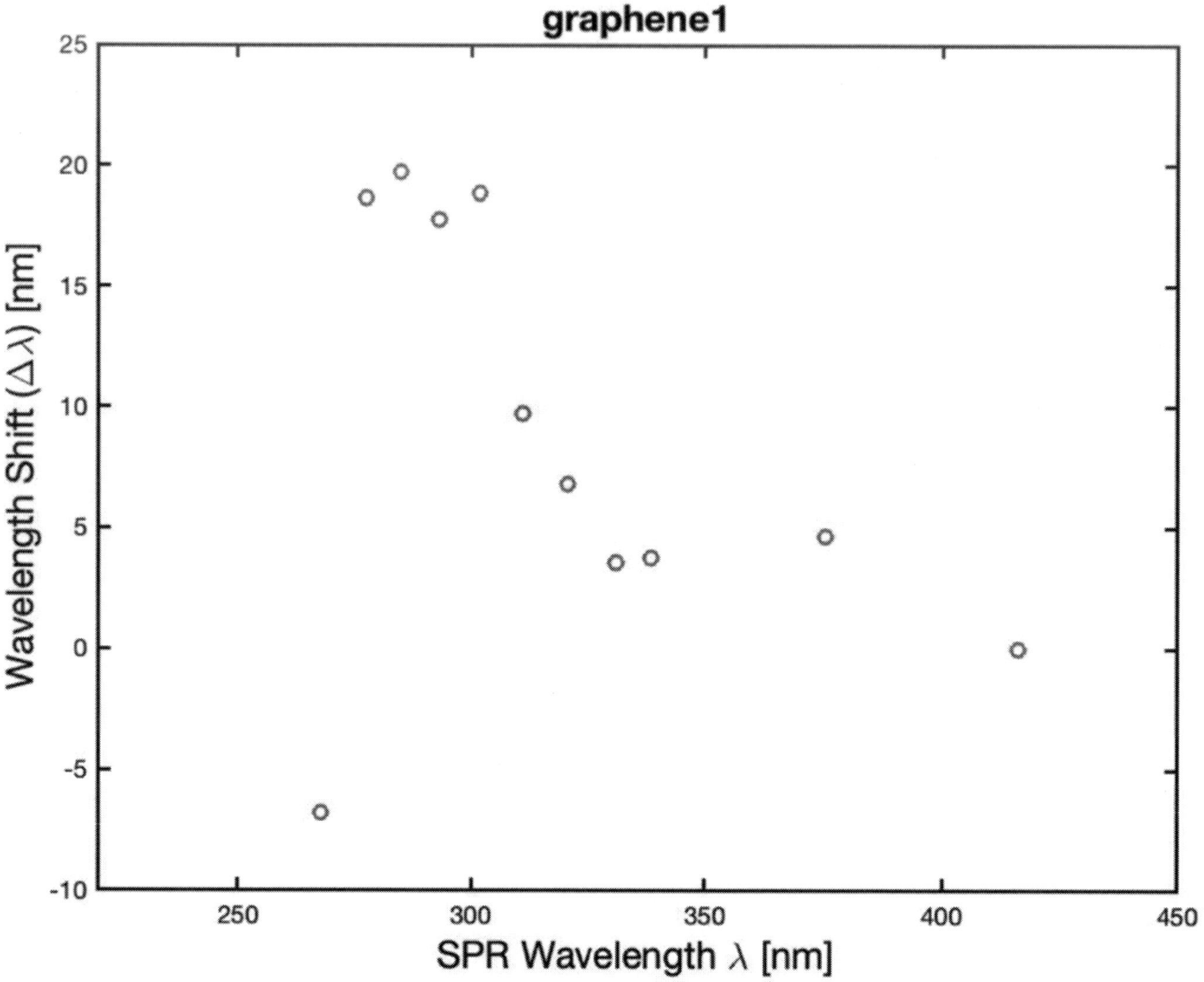

Figure 2. Wavelength shift after loading graphene

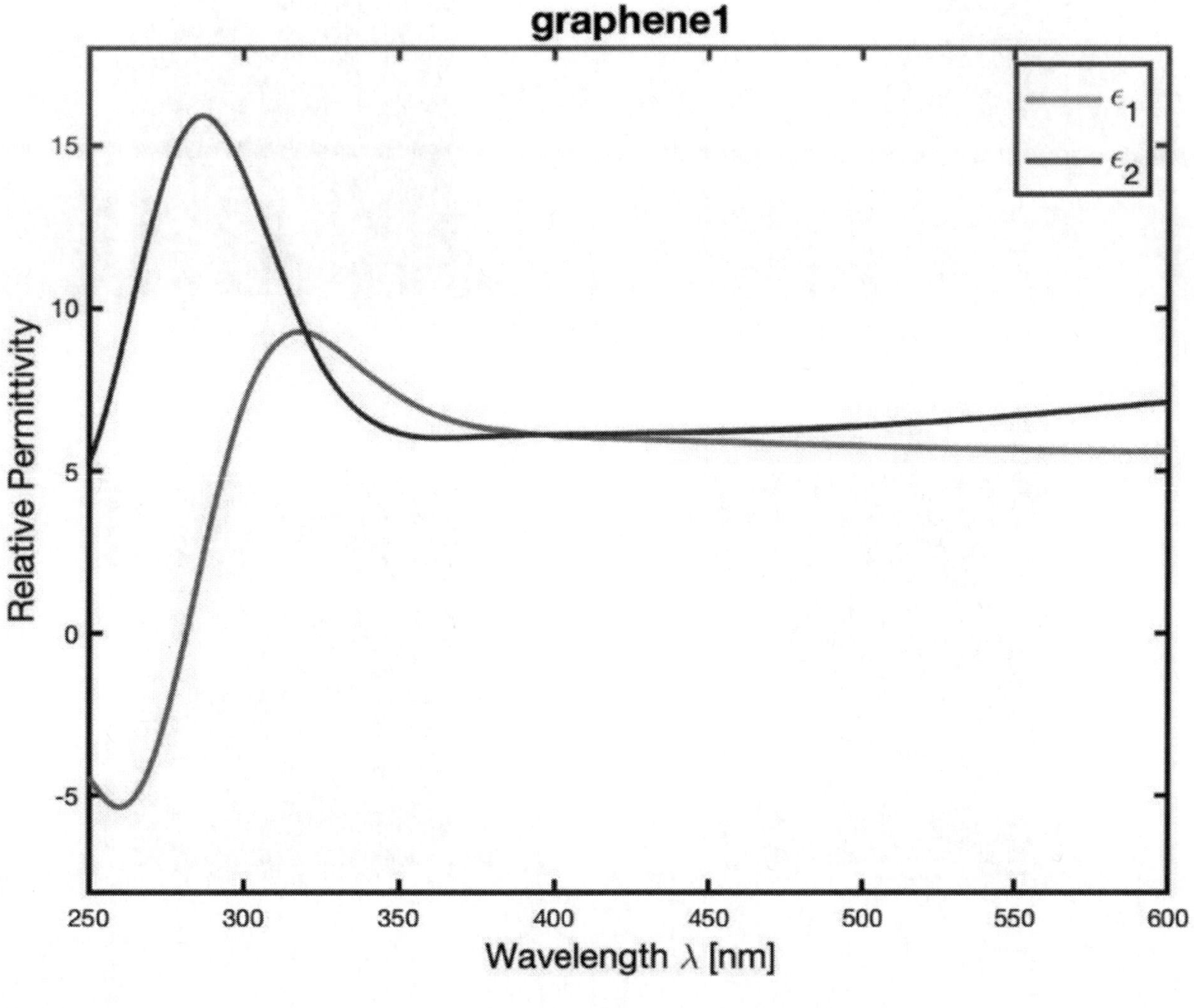

Figure 3. Permittivity of graphene

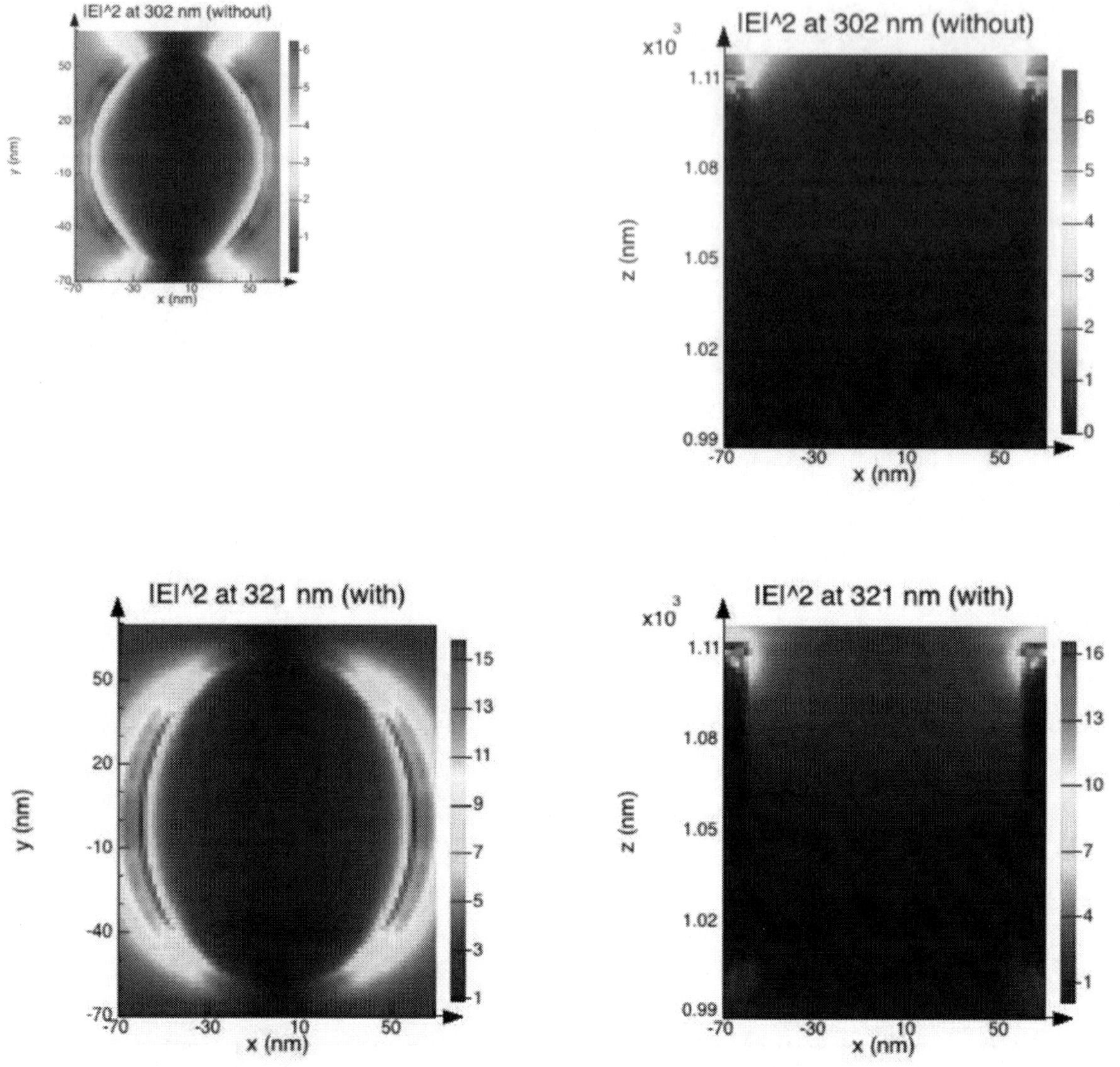

Figure 4. For hole array period equals to 280 nm, FDTD shows the near field intensity of the device without graphene (a) top surface and (b)cross-section, and the near field intensity of the device with graphene (c) top surface and (d) cross-section.

573766 Temperature Dependent Photovoltaic Characteristics of Graphene/h-BN/Si Solar Cell

Graphene-on-semiconductor heterojunction solar cell is an emerging class of photovoltaics with potential for efficient and reliable energy conversion systems. The interfaces between graphene and lightly-doped semiconductor play a key role in charge-carrier separation and recombination dynamics. Owing to the low Schottky barrier height-induced interfacial charge carrier recombination, the graphene-on-silicon (Si) heterojunction solar cells suffer from instability in power conversion efficiency over time. Therefore, it is critical to engineer the interface to enhance the barrier height by interfacing a chemically-stable, insulating, and atomically-thin layer. Further, the temperature dependent photovoltaic characteristics of such stacked architectures are unknown. Here, we have introduced hexagonal boron nitride (h-BN) as a tunneling interlayer in graphene-on-Si heterojunction solar cells, which enables the passivation of the chemical dangling bonds on the Si surface. The effect of temperature on the performance of graphene/h-BN/Si PV cell is also examined. Thin films of h-BN are directly synthesized on lightly-doped Si surface *via* a bottom-up chemical-surface-adsorption strategy followed by the transfer of a graphene monolayer. The 2D layer-on-2D layer-on-3D bulk semiconductor nanoarchitecture of graphene/h-BN/Si forms a metal-insulator-semiconductor (MIS)-type junction, where the h-BN acts as an electron-blocking layer to avoid interfacial charge carrier recombination. A 4-fold increase in open-circuit voltage (V_{OC}) is found for graphene/h-BN/Si heterojunction cell (0.52 V) in contrast to the graphene/Si cell (0.13 V), which is due to the increase in the Schottky barrier height and hence built-in electric potential. Interestingly, the V_{OC} linearly decreases by only ~4% with every 10 K increase in temperature. Durability studies indicate that the V_{OC} of graphene/h-BN/Si solar cell is practically unaffected for 1-year. This work will lead to an evolution of new 2D/2D/3D nanoarchitectures for mechanically-robust, high performance, and durable optoelectronic functionalities.

565563 Graphene-Based Passivation of Liquid Metal Particle Surfaces

Metals such as the eutectic gallium- indium (EGaIn) alloy that are liquid at room temperature are of particular interest in fields such as flexible electronics, self-healing circuitry, and soft robotics. They exhibit a unique combination of high thermal and electrical conductivity with mechanical softness but, in contrast to mercury, are nontoxic with low vapor pressures. However, these metal alloys are not without challenges. Trace amounts of oxygen will react with gallium and form an insulating skin layer, which can be a nuisance as it will adhere to most surfaces and leave behind undesirable residue. Gallium will also wet out and react with most other metal surfaces, even at room temperature, so is a poor choice for stable interconnects. The advancement of technologies utilizing these liquid metal materials hinges on the development of new strategies for managing these interfaces.

Graphene oxide (GO) and related materials have been shown previously to function as high performance interfacial micro-barriers in emulsified systems.[1] Here, we expand from previously studied organic/aqueous systems to both metal/aqueous and metal/organic particle dispersions, and are able to demonstrate the first encapsulation of EGaIn liquid metal particle cores by two-dimensional materials. Details discussed will include the role of the continuous solvent and graphene surface chemistries necessary to mediate the self-assembly of the graphene coatings.

1. Creighton, M. A.; Zhu, W.; van Krieken, F.; Petteruti, R. A.; Gao, H.; Hurt, R. H., Three-Dimensional Graphene-Based Microbarriers for Controlling Release and Reactivity in Colloidal Liquid Phases. *ACS Nano* **2016,** *10* (2), 2268-76.

571972 3D-Printed Graphene/Polymer Structures for Electron-Tunneling Based Devices

Designing 3D printed micro-architectures using electronic materials with well-understood electronic transport within such structures will potentially lead to accessible device fabrication for 'on-demand' applications. Here we show controlled nozzle-extrusion based 3D printing of a nano-composite of graphene/polylactic acid enabling the fabrication of a tensile gauge functioning via the readjustment of the electron-tunneling barrier width between conductive graphene-centers. The electronic transport in the graphene/polymer 3D printed structure exhibited a transport-barrier of 0.15 meV and a tunneling width of 0.79 – 0.95 nm (Fowler Nordheim electron tunneling) with graphene centers having a carrier concentration of $2.66 \times 10^{12}/cm^3$. Furthermore, a mechanical strain that increases the electron-tunneling width between graphene nanostructures (~5.5 nm) by only 0.16 Angstrom, reduces the electron flux by 1 $e/s/nm^2$ (from 18.51 to 19.51 $e/s/nm^2$) through the polylactic acid junctions in the 3D-printed heterostructure. We envision that the proposed electron-tunneling model for conductive 3D-printed structures with thermal expansion and external strain, will lead to an evolution in design of next-generation of 'on-demand' printed electronic and electromechanical devices.

558955 Cross-Linked Reduced Graphene Oxide Aerogels and Their Applications As 3D Electrodes for Lithium-Ion Batteries

Lithium-ion batteries (LIBs) are ubiquitous as they are the cornerstone of many electronic devices, such as smartphones, smartwatches, laptops, and power tools, due to their relatively low cost, good energy density, and slow self-discharge rates. While the battery cycling performance of conventional LIBs is satisfactory for some applications, challenges still exist to manufacture safe, light-weight, and more durable batteries to be used in places where a high performance-weight ratio is important, such as in aerospace and electric vehicles.

The energy density (energy per unit mass) of conventional LIBs is often limited by so-called electrochemically inactive materials, such as the binder, separator, current collectors, and battery casing. While increasing the mass loading/thickness of the active material (such as $LiCoO_2$ in the cathode of an iPhone battery) is a direct way to mitigate this loss in energy density, the Li-ion diffusion rate and the mechanical integrity of thick electrodes are often limiting. As such, the electrode thickness in conventional LIBs is restricted to ~100—125 μm. While developing novel active materials with greater theoretical capacity is important for improving the energy density of LIBs, such materials are often very expensive and can only be produced in labs, which is not favorable from a commercialization point of view.

Our work focuses on improving the energy density by substituting the metal current collectors with a three-dimensional, light-weight, mechanically robust, porous and conductive poly(acrylic acid)/reduced graphene oxide aerogel (further referred as rGO-XPAA aerogel). This aerogel has a bulk density of 6 mg/cm^3 and a porosity of 99.6%, and the thermally cross-linked poly(acrylic acid) (PAA) on the aerogel structure makes it elastic and mechanically robust.

Commercially available active materials, $LiNi_{1/3}Co_{1/3}Mn_{1/3}O_2$ (NCM) and $Li_4Ti_5O_{12}$ (LTO), were used to fabricate metal-free current collectors/electrodes with tunable mass loading (~ 3-30 mg/cm^2) and thickness (~ 30-130 μm). The electrodes can withstand ~ 100-fold compression (from ~3 mm to 30 μm) and achieve a high volumetric energy density of 1723 Wh/L for NCM and 625 Wh/L for LTO, which represents ~25% increase in energy density over their conventional counterparts.

579626 Investigation of Highly Nanoporous Metal Oxide Catalysts Via Nano-Template Synthesis for Oxygen Evolution Reaction

Over the decades, the generation of electricity using renewable energy such as photovoltaic (PV), wind, etc. is sorely expanded through the whole world by the national policies. For example, in Germany, PV-generated power covered approximately 7.2% (40 TWh) of the nation's net electricity consumption in 2017. As increasing the portion of renewable electricity, the role of power-to-gas (PtG) has been attracting great attention to store the excess electricity and to balance the grid, which can convert the electricity to hydrogen or methane. Recently, proton exchange membrane water electrolysis (PEMWE) has been developing due to the high current operation and excellent load-following characteristics compared to the alkaline water electrolysis for the PtG. Among the electrocatalytic reaction of the PEMWE, the oxygen evolution reaction (OER) at the anode displayed a significantly high overpotential. Thus, the catalysts for OER has been widely developed for several decades to improve the total efficiency of PEMWE. In this presentation, nanoporous RuO_x-based binary catalysts were synthesized by a nano-template method, using the ordered mesoporous silica material as a template and were investigated to improve the activity and stability for OER. Through this method, the nanoporous OER catalysts with a very high surface area over than 100 m^2/g were achieved. To investigate the effect of a secondary element in the RuO_x oxide, Ir and Sn were incorporated in the initial preparation stage. In half cell test, OER activity and stability were evaluated by an over-potential at 10mA/cm^2 and by using galvanic mode, respectively. Synthesized IrRuOx and SnRuOx catalysts displays both very small over-potential of 0.26 V at 10mA/cm^2, which indicates the addition of the second metal in the nanoporous RuOx catalyst is improved the activity towards for OER under acidic condition. These catalysts showed superior OER activity compared with the commercial RuO_2 sample

579253 Research and Application of Reinforced Composite Polymer Electrolyte Membrane in Fuel Cell for Transportation Applications

A polytetrafluoroethylene (PTFE)-reinforced perfluorosulfonic acid (PFSA) composite material is a cost effective alternative to the pure polymer electrolyte, which is often applied as polymer electrolyte membrane in fuel cell for transportation applications simply by infiltrating PFSA ionomer solution into porous PTFE matrix. For the membrane production, the reduction in cost and electrolyte thickness is an attractive factor, but it is still a challenge to produce the composite membranes as efficient as commercial membranes. Recent research and application activities with the reinforced composite membrane for the fuel cell applications include infiltration of the PFSA ionomer into the PTFE support, improving proton conductivity, reducing gas permeability, and addition of radical scavengers. In this talk, our recent results from these activities will be presented.

567671 Strain-Driven Exsolution in Perovskite Oxide for Improved Catalytic Activity at the Anode of Solid Oxide Fuel Cell

Redox ex-solution, *in-situ* synthesis process of metal nanoparticles upon high-temperature reduction has widely been studied as a way to fabricate metal nanocatalyst-decorated oxide electrodes for solid oxide electrochemical cells. However, the underlying mechanisms related to this phenomenon are not completely understood and the practical solutions that effectively accelerate B-site cation ex-solutions in perovskite oxide (ABO_3) have not yet been proposed. Here, we control the degree of Co enrichment at the surface of $SrTi_{0.75}Co_{0.25}O_{3-d}$ epitaxial thin films through lattice strain. Combined theoretical and experimental analyses show that the tensile strain that plays a role in stretching B-O bond length promotes the Co ex-solution in $SrTi_{0.75}Co_{0.25}O_{3-d}$. To practically make use of these findings, we introduce a Ba cation, which is a large-sized isovalent dopant instead of a Sr cation, to expand the B-O bond length in the perovskite lattice. We successfully demonstrate that this approach can facilitate the ex-solution of B-site cations, suggesting a new strategy for improving the surface reactivity of perovskite-type oxides.

568788 Effect of Side Chain Length in Phosphoric Acid Based Membranes on Nanophase-Segregation and Transport

Due to the high conductivity and stability, perfluorinated sulfonic acid (PFSA) proton exchange membranes such as Nafion are currently the standard for polymer electrolyte membranes (PEMs). While boasting superior conductive properties at standard conditions, Nafion dehydrates at high temperature and low humidity, drastically compromising its performance. While functionalization of PFSA membranes with phosphoric acid has been shown to improve conductivity under low-humidity conditions, these membranes still require optimization as they lack a practical level of conductivity. In this study, we investigate the nanophase-segregation and transport in phosphoric acid membranes in comparison to Nafion membranes. We also analyze the effect of side chain length on these two properties. To do this, we use density functional theory to prepare the monomeric structures for a sulfonic acid based polymer (Nafion) and a phosphoric acid based polymer. We then build 4 polymer chains with a degree of polymerization of 10 and construct a hydrated, charge-balanced simulation cell. The simulation cells are equilibrated using an annealing molecular dynamics (MD) simulation. After the equilibration, a long-run NPT simulation of 30 ns is performed to collect data for statistical analysis. We then assess nanophase-segregation through various gauges such as pair correlation functions and the scattering factor. Finally, we attempt to estimate how the chemical structures of monomer affect nanophase-segregation and transport. It is found that, as in sulfonic acid membranes, larger water clusters are formed in phosphoric acid membranes with longer side chains. However, unlike in sulfonic acid membranes, the phosphoric acid groups associate closely with one another via hydronium bridging, which suppresses proton hopping. Therefore, it is inferred that the proton transport is decreased with increasing the phosphoric acid side chain length. Based on these results, we recommend opting for shorter side chain lengths when designing phosphoric acid membranes.

570518 Quantitative Study of the Stability of Colloidal Metallic Nanoparticles with Different Shapes in Reactive Chemical Environments at Atomic Level

Understanding and controlling electrochemical interfaces at atomic and molecular levels have transformed electrochemistry into a science with clearly defined fundamental principles leading to significant impact on various electrochemical systems and devices. Although the principles guiding the activity of electrochemical reactions are quite well established, the driving forces that control stability are still poorly understood. Here we utilize in situ monitoring of the early stages of Pt, Pd, and Ni dissolution using the operando attenuated total reflectance surface enhanced infrared absorption spectroscopy (ATR-SEIRAS). Our unique operando ATR-SEIRAS method provides femtogram sensitivity levels that, in combination with other techniques, provide otherwise inaccessible information about the dissolution of colloidal metallic nanoparticles with different shapes in acidic environments, and thus deliver a more sensitive approach to understanding the structure-stability relationships of colloidal metallic nanocatalysts at atomic level.

577988 Facile One Pot Synthesis of Palladium-Based Alloy Wavy Nanowires for Hydrogen Separation

Palladium and palladium alloy nanowires (PdM; M= Ag, Cu, Y) of varying composition were synthesized using a facile and scalable one-pot polyol reduction of metallic salts using poly(vinylpyrrolidone) (PVP) as a capping agent. The alloy nanowires with high surface area and interconnected dendritic network structure could provide efficient transport pathways for hydrogen through a membrane. Incorporation of an appropriate amount of a metal like Ag, Cu, or Y can increase the permeability of hydrogen through the metal while reducing the usage of palladium. Mixed-matrix membranes containing the palladium alloy nanowires dispersed in poly(benzimidazole) (PBI) polymer and their separation performance was tested at 150 C and 100 psig for separation of hydrogen from a 1:1 mixture of H_2 and CO_2. Initial test results show increased permeability of H_2 up to 52.7 barrer and with pure PBI membrane containing 50 weight percent $Pd_{0.76}Ag_{0.24}$ alloy nanowires 43.9 barrer as compared to just 25 barrer of pure PBI membrane.

575905 (Invited Plenary Talk): Reproducibility and Reusability of Surface Bound Electrochemical Sensors

Polymer structures on sensor surfaces are known to have reproducibility and reusability issues caused by hydrodynamic shifts, minor changes in surface coverage, charging of the surface, and steric restructuring of the polymer surfaces. Within our laboratory, we primarily use elastin-like polymers and hydrophobic-self-assembled-monolayer:cyclodextrin complexes as functional polymers on sensor surfaces. Often, with these polymers, the same methodology does not lead to the same surface modification due to small minor surface changes including physiosorbed surface content or steric hinderances. Therefore, attention and strategy are needed to ensure a reproducible sensor surface. Further, we use electrochemical impedance spectroscopy, a highly sensitive but difficult to reproduce strategy in biosensors because small surface changes can result in large changes in the recorded signal. We will discuss our methodological approaches to achieve reproducible and reusable sensor using polymer surfaces with electrochemical impedance spectroscopy. Our group cares about reusability as a long-term sensor strategy towards measuring bioanalyte content *ex vivo* in biofluid samples. We will discuss both positive and negative approaches to generate the same results over multiple sensors, and the ability to regenerate a sensor for reuse. Further, we will talk about voltage dependency of the surface and ability to reproducibly detect nanomolar chemical changes.

The author would like to acknowledge the support from NSF EAGER (CBET 1638896) and NIH CIBBR (P20GM113131). This contribution would not be possible without the endless support of the undergraduate students, graduate students, post doctorates, and collaborators of the Surface Enhanced Electrochemical Diagnostic Sensors Laboratory.

576041 (Invited Plenary Talk): Electrochemical Sensors for Multiplexed Infection Detection and Monitoring

Bacterial cells secrete hundreds of different molecules as part of their natural processes. These molecules range from quorum sensing molecules or pheromones for communication, to siderophores as nutrient scavengers, to virulence factors that attack intruders. The ability to quantitatively measure the concentrations of these molecules is important for both research and clinical applications. This talk will describe electrochemical sensors that are being developed and used for diagnosing infection and monitoring bacterial physiology. The sensors utilize voltammetry to measure interactions of redox-active molecules with electrode surfaces as a signal transduction mechanism to track binding of target molecules with recognition elements. The method works directly in biological fluids, allowing ex vivo measurements to be easily performed.

569320 (Invited Plenary Talk): Hybrid Plasmonic Nanomaterials for Uranium Sensing

Uranium, a radioactive material with a long half-life, accumulates in the environment in its oxidative form uranyl, which can contaminate soil and water. Importantly, uranyl forms complexes with anions and cations thereby influencing solubility, toxicity, and fate of these heavy metal species. Furthermore, uranyl speciation varies with pH. The resulting complex speciation complicates detection and/or requires significant sample pretreatment. As such, methods that are capable of identifying trace uranyl species in complex samples are needed. In this presentation, hybrid plasmonic nanomaterials and experimental workflow implementation will be used to establish a rigorous protocol for uranyl sensing and detection. Namely, localized surface plasmon resonance spectroscopy (LSPR), Raman spectroscopy, and surface enhanced Raman scattering (SERS) serve as label-free and near real-time methods for identifying uranium species in complex aqueous solutions. Reproducible detection of uranyl, an important biological and environmental contaminant, from complex matrices is successfully achieved using hybrid plasmonic nanoparticles. Traditionally, non-specific binding of interfering species limits detects of molecules such as uranyl. Herein, this is overcome using materials design and rigorous sample analysis workflow design. Synergistic approaches for uranyl isolation and SERS detection is promising for real-world sample detection and eliminates the need of radioactive tracers and extensive sample pretreatment steps.

579768 (Invited Plenary Talk): Continuous, Real-Time, Physiological Monitoring with Nanosensors

Continuous physiologic and therapeutic monitoring has the potential to redefine prediction, diagnosis, and treatment of disease. Current approaches focus on intermittent blood sampling and analysis – an approach that is not scalable or predictive with decisions based on cohort values rather than individual history. Future approaches will draw on continuous monitoring to track metabolic parameters before, during, and after treatment. This enables real time feedback, as well as comparison to a patient's own historical trends and fluctuations. Our nanosensor platform has the ability to continuously monitor analyte levels with straightforward optical measurements. We, and others, have developed sensors for a wide variety of ionic analytes (e.g. sodium, potassium, lithium, pH) as well as nonionic analytes (e.g. glucose, histamine) which are relevant for monitoring in clinical and research settings. In this presentation I will discuss design and fabrication of nanosensors, control of sensor response, incorporation of next generation fluorescent tools into our sensors, and applications of this platform for physiological monitoring and the study of biofilm metabolism. This sensor approach shows potential as a research or medical tool for continuous monitoring of drug and metabolite concentrations without the need for direct blood samples.

563453 Purification of Single-Wall Carbon Nanotubes Using Aqueous Two Phase Extraction to Single (n,m) Species in the Large Diameter Limit

Separation of single-wall carbon nanotube (SWCNT) populations to pure specified physical structures is a primary milestone necessary for enabling SWCNT-based technologies. Aqueous two-phase extraction (ATPE), in which the spontaneous phase separation of an aqueous polymer mixture is utilized to spatially separate solutes on the basis of their chemical potential difference for the two phases, is a promising technology for achieving this milestone. However, due to the expanded number of (n,m) species with increasing nanotube diameter, and the increased difficulty in optical characterization of aqueous-dispersed samples, isolation of larger diameter SWCNTs, > ~1.4 nm in diameter, has continued to be a difficult proposition. Advances in methodology of separation and sample preparation are now overcoming both these barriers in our laboratories. In this work I will detail improved separation procedures utilizing alkane-filled SWCNT populations for the isolation of individual SWCNT (n,m) structures from > 1.5 nm average diameter SWCNT dispersions. These method developments can readily isolate multiple large diameter (n,m) species, and are enabling new scientific and device application work. The enhanced separation methods, isolated (n,m) species, and applications of those populations will be presented.

554610 Utilizing a Carbon Nanomaterial-Based Nanocomposite Aerogel Towards Separation and Energy Applications

Overutilization of organic-based compounds (e.g. fertilizers) in agricultural settings continues to cause concern due to potential adverse impacts on the health of humans and the environment. Adsorbents, such as activated carbon, are widely used due to their ability to adsorb various contaminants. Limitations in the effectiveness of activated carbon, where high uptake capacities are desired, require the need for alternatives to current state-of-the-art adsorbents. In this work, the adsorption capability of a lightweight, carbon-based aerogel is presented. The aerogel is composed of 0 to 2 wt% mixture of both graphene nanoplatelets (GnP) and single-walled carbon nanotubes (SWCNT). Two organic compounds, 2,4-dichlorophenoxyacetic acid (2,4-D) and 1-pyrenebutyric acid (PBA), served as the contaminants of interest. Results show that the aerogel containing 0.2 wt% SWCNT-GnP achieved an uptake capacity of 0.22 mg 2,4-D/mg aerogel and 0.083 mg PBA/mg aerogel. Compared to activated carbon, from a kinetic standpoint, the aerogels produced in this work demonstrated 39 times higher adsorption of 2,4-D and 5 times higher adsorption of PBA making them a viable candidate as a next generation adsorbent. In addition to separation applications, these carbon nanomaterial-based aerogels have also been studied for use in energy applications (e.g. Na-ion batteries and supercapacitors) and will be discussed.

576325 Study of Unique Kinetic Vs Thermodynamic Equilibria of Specific Chiralities of Single Walled Carbon Nanotubes

Study of Unique Kinetic vs Thermodynamic Equilibria of Specific Chiralities of Single Walled Carbon Nanotubes

Aniruddha Kulkarni, Stephen Michel, Dr. Yang Zhao and Dr. Kirk J. Ziegler

University of Florida, Gainesville, USA.

kziegler@che.ufl.edu

Single walled carbon nanotubes (SWCNTs) find applications in electronics and biosensors owing to their unique optical properties. The range of applications is limited due to constraints in separating highly pure monochiral fractions. It has been reported that surfactant structure around the nanotubes plays an important role in separating pure fractions of desired (n,m) species. The presence of thermodynamically stable co-surfactant states around single walled carbon nanotubes enables the selective coating of (n,m) species in the solution. The desorption of surfactants is based on formation of these states at very specific ratios. Our group has previously utilized this approach to achieve selective elution of (6,5) nanotubes at very specific critical exchange ratio concentrations (CERC). The co-surfactant solution utilizes a ratio of sodium dodecyl sulphate (SDS)/sodium deoxycholate (DOC) to obtain single-chiral eluents of (6,5) species.

The exchange of surfactants on the surface of the nanotubes results in different kinetic and thermodynamic states. Adsorption of surfactant occurring at varying time scales based on the elution time differences suggests kinetically driven exchange. Here, we present unique behavior between kinetically and thermodynamically driven exchange of DOC on SDS coated SWCNTs. The solvatochromatic shifts of (n,m) species attributed to these exchanges taking place on the nanotube surface are probed with optical spectroscopy. The nature and extent of these shifts are based on chirality at specific CERC. Equilibrium studies were performed on specific chiralities to obtain thermodynamic parameters. These findings provide a promising pathway for investigating the possible role of enantiomers in coating of DOC to nanotubes.

570027 Transformation of GO Surfaces to Reduced-GO Surfaces with Tunable Surface Chemistry and Morphology Via Pool Boiling Process

Owing to supreme thermal conductivity and wetting properties, graphene coatings have been exploited in numerous two-phase heat transfer applications such as electronics cooling, reboilers, refrigeration, and desalination. Boiling is a heat transfer process accompanied by phase change from liquid to vapor, subsequently, pool boiling involves the boiling of a stagnant liquid over a heated surface. For various manifestations, the pool boiling performance of a surface is dictated by higher critical heat fluxes and heat transfer coefficients. This talk will focus on use of repetitive pool boiling as a tool to test the durability of graphene oxide-copper (GO-Cu) coated surfaces and the subsequent transformation of GO-Cu to reduced GO-copper (rGO-Cu) coating that further led to improved boiling performance. This improvement is attributed to a) transition of hydrophilic GO to hydrophobic rGO due to the repetitive pool boiling, and b) increased surface roughness because of hydrophobic surfaces. The durability of the GO-Cu coatings was found to be higher than pristine-copper and just-copper coated surfaces. A higher performance demonstrated by GO-Cu coatings indicates its increased sustainability and practicability in numerous engineering applications.

560227 Harnessing the Intrinsic Synthesis Catalysts to Create Faradaic Carbon Nanotube Composites

The development of new materials to improve interfacial charge transfer characteristics will drastically enhance energy storage, heterogeneous catalysis, and many other electrochemical applications. Here we report two simple procedures that can harness the Faradaic nature of residual iron nanoparticle catalysts that endure within Multi-walled Carbon Nanotubes (MWNT) post-synthesis, thereby alleviating the challenges associated with forging hybrid nanocomposite electrodes. Non-purified MWNTs, undergo a chemical oxidation process in acidic conditions with KMnO4 to partially "upzip" the MWNTs and exposes the redox-active iron nanoparticles to the bulk liquid. Electrochemical activation of these non-purified MWNTs is also possible by exposing the tubes to high voltages in acidic conditions causing the tubes to rupture partially and permitting the interaction between the catalysts and the electrolytes. In both cases a stable redox peak associated with the $Fe^{2+/3+}$ transition is achieved during the MWNT oxidation process yielding a ~350% increase in capacitance (>300 F/g) relative to purified MWNT electrodes (70 F/g). While these materials solely may be pragmatic as energy storage electrodes, the integration of redox species within an inert carbon electrode will also provide new opportunities to accelerate heterogeneous charge transfer reactions.

577156 Packaging Vertically Aligned Carbon Nanotubes into a Heat Shrink Tubing for Efficient Removal of Phenolic Pollutants

Having extremely high surface-to-volume ratio, carbon nanotubes (CNTs) are excellent adsorbents for the removal of organic pollutants. Majority of studies utilizing CNTs for such purpose disperse CNTs in a batch-type reactor filled with a solution of pollutants. However, retrieval or collection of CNTs after the adsorption can often be more demanding than the removal of pollutants. In this study, we address the challenge by packaging vertically aligned CNTs into a PTFE heat shrink tubing. Insertion of CNTs into the tubing and its thermal shrinkage densified the CNTs radially by 35%, which also reduced wrinkles in the nanotubes as confirmed by scanning electron microscopy (SEM). The CNT-based adsorption tube with a circular cross-section enabled facile connection to a source of polluted water, which we demonstrated by connecting the tubular platform to a syringe filled with solution of phenolic compounds. We passed the solution through the adsorption tube and analysed the filtrate in gas chromatography, which gave adsorption capacity of 72, 85, and 66 mg/g for phenol, m-cresol, 2-chlorophenol, respectively. Another advantage of our platform is that CNTs can be easily functionalized by flowing reactants through the tubing. Functionalization of CNTs with carboxyl groups, by flowing solution of nitric acid, enhanced the corresponding adsorption capacity to 115, 124, and 81 mg/g. We attribute the high adsorption capacity of our platform to efficient adsorbate-CNT interaction within the narrow interstitial channels between the aligned nanotubes. The CNT-based adsorption tubes can be of great significance for the simple and efficient removal of phenolic and other types of organic pollutants.

572220 Efficient and Mechanically Robust All-Polymer Solar Cells

Non-fullerene acceptors (NFAs), that are small-molecule acceptors (SMA) and polymer acceptors (PAs), have been extensively explored, which has yielded significant enhancements in the photovoltaic performance of polymer solar cells (PSCs). The mechanical robustness of the PSCs is of vital and equal importance to ensure the long-term stability and enable their use as power-generators in flexible and stretchable electronics. Here, we report a comparative study of the mechanical properties of SMA-based, PA-based, and fullerene-based PSCs. We chose ITIC (SMA), P(NDI2OD-T2) (PA), and PCBM (fullerene) as three representative acceptor materials and blended them with the same polymer donor PTB7-Th. To understand the difference between the mechanical properties of SMA-based and PA-based PSCs, we control the number-average molecular weight (M_n) of P(NDI2OD-T2) from 15 to 163 kg mol^{-1} in all-PSCs. The all-PSCs based high M_n PA exhibit significantly higher cohesion energy (4.03 J m^{-2}) than SMA-PSCs (1.19 J m^{-2}) and PCBM-PSCs (0.29 J m^{-2}). Notably, the all-PSCs exhibited a high strain at fracture of 31.1%, which is 9- and 28-fold higher than those of SMA-PSCs and PCBM-PSCs, respectively. The superior mechanical robustness of all-PSCs is attributed to using a PA above the critical molecular weight (M_c), which produces tie molecules and polymer entanglements that dissipate substantial mechanical strain energy with large plastic deformation. This work provides useful design guidelines for photovoltaic active materials in terms of the mechanical properties and highlights the importance of incorporating high M_n PAs above the M_c for producing the PSCs with excellent mechanical robustness and device performance.

572889 Factors Governing Site Specific Rates of Electronic Excitations in Multicomponent Plasmonic Materials

The decay of localized surface plasmons supported by plasmonic metal nanoparticles results in the formation of energetic charge carriers within the nanoparticles. These energetic charge carriers can then be extracted from the plasmonic material and transferred to other materials (e.g. metals, semiconductors, or molecules) that can then perform a function. The efficient extraction and exploitation of these charge carriers in various applications can be significantly improved by designing multicomponent plasmonic nanostructures with highly localized charge carrier generation rates in the non-plasmonic component. In this contribution, we use experimental and computational studies of core–shell metal–metal, metal–semiconductor and metal–molecule systems to investigate the mechanism of energetic charge carrier generation in multicomponent plasmonic systems [1,2,3]. We reveal that the rates of plasmon decay through the generation of energetic charge carriers are governed by two factors: (1) the intensity of the confined plasmon induced electric fields at the surface of the plasmonic nanostructure, and (2) the availability of direct, momentum conserved electronic excitations in the material located where the electric fields are high. We use these studies to propose a unifying physical framework that leads us towards molecular control of excited charge carrier generation in all multicomponent plasmonic systems.

References

1. Aslam, U., **Chavez, S**. & Linic, S., *Nature Nanotechnol.* **12,** 1000–1005 (2017).
2. **Chavez, S**., Aslam, U., & Linic, S., *ACS Energy Letters*. **3,** 1590-1596 (2018).
3. **Chavez, S**., Govind Rao, V., & Linic, S., *Faraday Discussions*. Pre-print. (2019).

573427 Advanced Li-Metal Batteries and Their Thermal Safety Aspects

Advanced Lithium metal batteries experiences inherent concerns associated with poor Coulombic efficiency and the dendritic Li growth during repeated cycling. To significantly improve the cycle performance and electrochemical stabilities of Li metal electrodes, we designed multifunctional tri-layer membrane with polydopamine and a graphene-carboxymethyl cellulose on top of the standard polypropylene separator. The modified trilayer separator has an excellent electrolyte wettability, enhanced electrical conductivity and an additional capacity for Li storage. It ease the electrochemical reactions and kinetics in both the Li/Cu half-cell and the Li/$LiFePO_4$ full cells. The temperature dependent electrochemical performance and its thermal safety features will be discussed in great details.

567130 Quasi-2D Hybrid Organic-Inorganic Perovskites: DFT Modeling Approach

Hybrid organic-inorganic perovskites (HOIPs) have developed to become highly promising solution-processable direct-bandgap semiconductors, which have exhibited high solar conversion efficiencies due to their outstanding charge carrier mobilities. HOIPs exhibit superior optical properties due to their relatively high photoluminescence quantum efficiency and high brightness emission. Furthermore, HOIPs have garnered much interest due to their tunable bandgaps; which are controlled by modifying the dimensionality and composition of the perovskites. In this study, we utilize DFT methods to investigate the emission of HOIPs in order to achieve reliable tunability of bandgap. We investigate quasi-2D perovskites which are considered to form multiple quantum wells, in which the inorganic layers act as wells and the bulky organic spacers as barriers. It is well known that the bandgap of these layered perovskites can be tuned by varying the "quantum well" thickness (corresponding to *n* in $A'_2A_{n-1}B_nX_{3n+1}$). While it has been shown that the size of the organic cations occupying the cuboctahedral cavity of 3D perovskites indirectly affect bandgap by altering B-X orbital overlap through octahedral distortions, the effect of the spacer size (in the interlayer region) has not been fully understood in 2D perovskites. In this study, we investigate the effects of aliphatic and aromatic spacers on the bandgap and formation energies of HOIPs. Furthermore, we characterize a correlation between the steric size of the spacer cation and band gap, which is a key characteristic for the design of optoelectronics. Overall, we find that the choice of the spacer cation can play a crucial role in tuning the emission of perovskite light-emitting diodes.

562681 Carrier Dynamics of Photocatalytic Two-Dimensional Semiconductors Decorated with Plasmonic Nanoparticles

Layered transition metal dichalcogenides (TMDs), MX_2 structures where M = Mo or W and X = S, Se, or Te, are an emergent source of 2D semiconductors with intriguing catalytic and optoelectronic properties. Concurrently, plasmonic metal nanoparticles (NPs), featuring an extraordinary ability to concentrate light and energy in sub-wavelength dimensions, can enhance light harvesting and catalytic activity in adjacent semiconductors. Mechanisms of plasmonic enhancements includes local heating, light scattering, local field confinement, resonant energy transfer, and interfacial charge transfer. However, these underlying physical mechanisms are poorly understood *in situ*, and most devices have exhibited poor performance due to the localized nature of plasmon resonances and their picosecond carrier relaxation. The work herein utilizes transient absorption spectroscopy, with pump tunability and broadband visible-NIR probing, to monitor the carrier excitation and dynamics in AuNP-decorated TMDs. A combination of this time-resolved and steady-state optical spectroscopies unveiled the presence of a sub-gap absorption feature in gold NP-decorated molybdenum diselenide ($MoSe_2$). Optical excitation of this low-energy feature generates long-lived, charge carriers in $MoSe_2$. Improved understanding of interaction between plasmonic NPs and adjacent 2D semiconductors represents an important advance in the design and implementation of these emerging hybrid materials in solar photoelectrochemistry.

579671 Development of Tandem Semiconductor Microparticles for Water-Splitting

There has been a vast improvement in photovoltaic (PV) technology in the last couple of decades; however, solar PV technology has a lot of drawbacks, including that it is intermittent, broadly dispersed and non-transportable. Hence, cost-effective solar energy storage is critical for widespread implementation of solar energy as the primary energy source. My project focuses on development of tandem semiconductor microparticles which, when placed in water under 1 Sun condition, will spontaneously split water into hydrogen and oxygen with solar-to-hydrogen efficiency of >10%. In the present work, we have acquired customized microwire array structures made on silicon wafer by deep reactive ion etching method and formed a diode on these particles via a doping process. We have also been able to produce a tandem junction for additional photovoltage. Some oxygen evolution catalyst was then deposited on the top subshell of the junction and the current-voltage characteristics was measured of this semiconductor assembly for oxygen evolution reaction.

Future work aims at detaching these particles from the Si substrate and suspending them in a metal precursor solution for depositing the hydrogen evolution catalyst. These particles will then be suspended in the electrolyte under 1 Sun condition and the evolved hydrogen will be measured by Gas Chromatography.

570351 Synthetic-Biological Hybrid Nanoscale Systems for Drug Delivery

Nanoparticle-based drug delivery systems are widely explored to improve the biological outcome of encapsulated drugs including small molecules, peptides and nucleic acids. However, poor vascular circulation, limited targeting and the inability to negotiate many biological barriers are key hurdles in their clinical translation. Improved targeting and reduced off-target accumulation is essential to extract maximal benefits from nanoparticle systems. Biology has provided many examples of successful "carriers" in the form of circulatory cells, which routinely overcome the hurdles faced by synthetic nanoparticle systems. Our laboratory has explored blood-cell inspired drug delivery systems that take advantage of the abilities of red blood cells and monocytes. We have explored "cellular hitchhiking" which involves combining synthetic particles with circulatory cells to drastically alter the *in vivo* fate of the synthetic particles. I will provide an overview of the principles and capabilities of these nanoparticle drug delivery systems.

570250 Engineering Muco-Inert Nanoparticles for Improved Vaginal Drug Delivery during Pregnancy

Introduction: Vaginal drug delivery is the preferred route for targeting the female reproductive tract. By avoiding the harsh gastrointestinal environment and hepatic first-pass effect, local tissue concentration can be increased [1]. Additionally, the uterine-first-pass effect leads to accumulation of vaginally administered drugs in the uterus [2]. Vaginal formulations have been demonstrated to have some efficacy in prevention of preterm birth (PTB), though the optimal formulation for vaginal drug delivery has yet to be explored. Key barriers to effective vaginal drug delivery, including the protective mucus that acts as a barrier and facilitates clearance, must be considered [3]. By identifying physicochemical properties for optimal nanoparticle penetration of CVM, improved vaginal drug absorption can be achieved. Here, we characterized the structural and barrier properties of cervicovaginal mucus (CVM) during pregnancy using fluorescent probe nanoparticles of various sizes and surface coatings. We further utilize these design criteria to formulate drug nanoparticles that can penetrate the vaginal mucus barrier for more effective prevention of PTB.

Methods: 500 CVM samples (spaced monthly) were self-collected by 100 pregnant women. CVM microstructure was characterized by a quantitative multiple particle tracking method. Fluorescently-labeled nanoparticles with and without mucoinert coatings were added to CVM. Particle motions due to thermal energy were recorded and quantified to estimate the mesh spacing of mucin proteins comprising CVM. Further, these nanoparticles were administered vaginally to pregnant mice and the *in vivo* distribution was imaged. Several drugs were formulated into nanosuspensions via wet milling. The size and zeta potential of the formulations were measured. Drug levels the female reproductive tract tissues were analyzed using LC-MS/MS methods. Finally, these formulations were tested in a murine model of PTB. Pregnant mice were challenged with an intra-uterine (IU) lipopolysaccharide (LPS) injection on embryonic day 15 (E15, out of 19 day gestation). Treatment was dosed vaginally daily from E15-18. Mice were checked every

Results: Similar to what was observed previously in CVM from non-pregnant women, nanoparticles with mucoinert coatings have increased mobility in CVM from pregnant women. However, CVM from pregnant women causes a significant reduction in mobility of nanoparticles, particularly for nanoparticles 500nm in size or greater, as pregnancy progresses. Mucoinert nanoparticles ~250nm in size were slowed, but a significant fraction were still highly mobile in CVM. This inferred reduction in pore size in CVM is likely due to increasing progesterone levels throughout pregnancy. Pharmacokinetic studies revealed that nanosuspensions with a size <300 nm and a zeta potential between 0 mV to -5 mV were better able to deliver drugs to the female reproductive tract tissues. Additionally, vaginally delivered drugs were shown to target the female reproductive tract as compared to systemically administered formulations. We were able to formulate Trichostatin A, progesterone, hydroxyprogesterone caproate, vorinostat, and several other drugs into nanosuspensions. Progesterone and its analogs are prescribed for the prevention of PTB clinically. Surprisingly, progestins were only able to prevent PTB in our model when in the presence of a histone deacetylase inhibitor. Analysis of cervical and myometrial tissues revealed anti-inflammatory effects from the combination dosing.

Conclusion: Here, we gained valuable insight for engineering targeted vaginal drug treatments for pregnant women. We have demonstrated the first prevention of preterm birth in this animal model, and are able to bring the dams to deliver live pups. Effectively delivering drugs to target tissues can help to reveal mechanism behind disease. Our technology has allowed us to better target the female reproductive tract and understand mechanisms of preterm birth.

1. Ensign, R. Cone, J. Hanes, JCR 2014, 0:500-5:14
2. De Ziegler, C. Bulletti, B. De Monstier, A.S. Jääskeläinen, Ann N Y Acad Sci, 1997; 828:291-9.
3. Yu, J. Chisholm, W.J. Choi, A. Anonuevo, S. Pulicare, W. Zhong, M. Chen, C. Fridley, S.K. Lai, L.M. Ensign, J.S. Suk, J. Hanes, Adv Healthc Mater, 5 (2016) 2745-2750.

4. J.C. Condon, P. Jeyasuria, J.M. Faust, J.W. Wilson, C.R. Mendelson, PNAS 2003, 100.16, 9518-9523.

573858 Toward a Stable Single-Dose Oral Malaria Cure: Enhancing Dissolution Kinetics Using Continuous Nanoparticle Fabrication

AIChE 2019 annual conference abstract

"Toward a stable single-dose oral malaria cure: enhancing dissolution kinetics using continuous nanoparticle fabrication"

Kurt Ristroph, Jie Feng, Simon A. McManus, Yingyue Zhang, Kai Gong, Hanu Ramachandruni, Claire E. White, and Robert K. Prud'homme

OZ439 is a new chemical entity which is active against drug-resistant malaria and shows potential as a single-dose cure. However, development of an oral formulation with good bioavailability has proved problematic, as OZ439 is a poorly soluble (BCS Class II) drug. In order to be feasible for low and middle income countries, any process to formulate such a therapeutic must be inexpensive at scale, and the resulting formulation must survive without refrigeration even in hot, humid climates. We here demonstrate the scalability and stability of a nanoparticle formulation of OZ439. We apply a combination of hydrophobic ion pairing and Flash NanoPrecipitation (FNP) to formulate OZ439 NPs 150 nm in diameter using the inexpensive stabilizer hydroxypropyl methylcellulose acetate succinate (HPMCAS). The nanoparticles are then processed into a dry powder by spray drying. Both FNP and spray drying are continuous unit operations that can be run continuously and inexpensively at scale.

The spray dried NP powders achieve over 20-fold higher OZ439 solubilization than unprocessed drug in vitro in simulated intestinal fluid. The powders are then subjected to a series of accelerated aging stability trials at high temperature and humidity conditions. The dissolution kinetics of OZ439 in the NP powders remain constant across 1 month in uncapped vials in an oven at 50°C and 75% RH, and for 6 months in capped vials at 40°C and 75% RH. X-ray powder diffraction and differential scanning calorimetry profiles similarly remain constant over these periods.

The combined nanofabrication and drying process described herein is an important step toward an industrially-relevant method of formulating the antimalarial OZ439 into a fast-dissolving single-dose oral form with good stability against humidity and temperature.

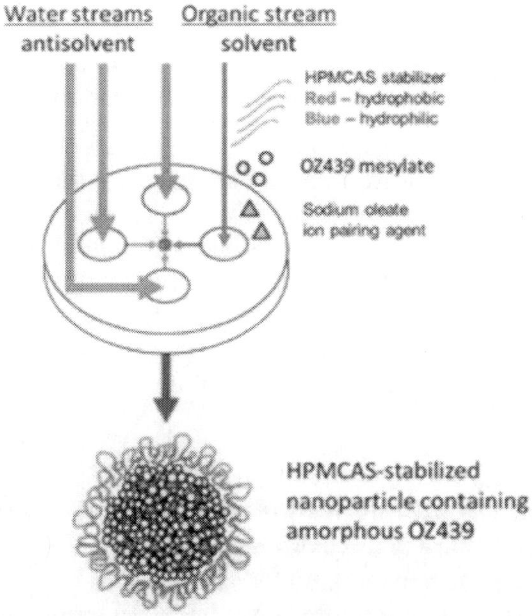

Schematic of Flash NanoPrecipitation (FNP) used to ionically complex the poorly-soluble antimalarial OZ439 and encapsulate it into a nanoparticle. This schematic depicts an intermediate-scale multi-inlet vortex mixer that is operated at 160mL/min.

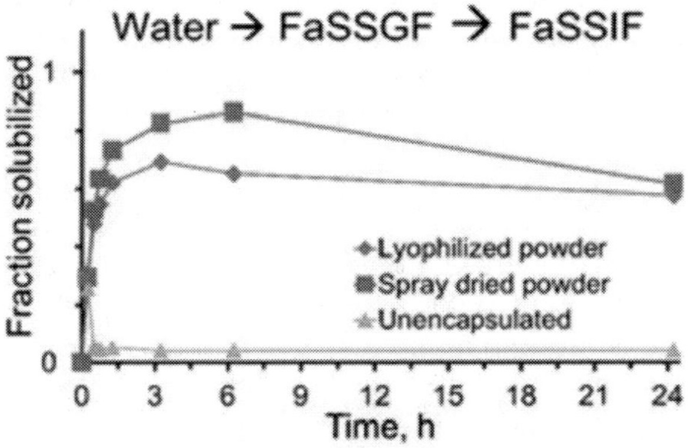

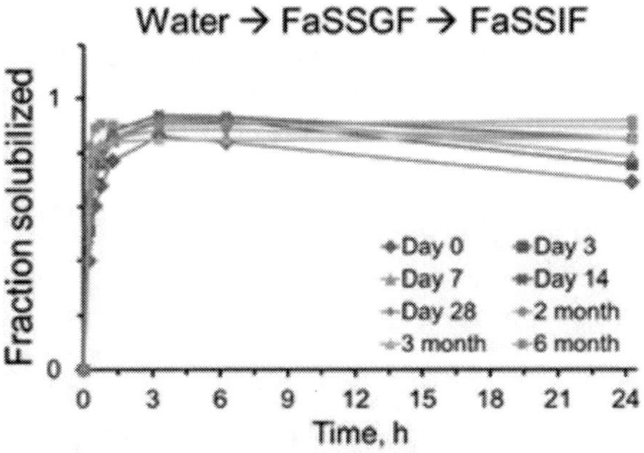

Left: Dissolution kinetics of spray dried NPs, lyophilized NPs, and bulk drug *in vitro* in fasted-state intestinal fluid. The spray dried powders exhibit the fastest OZ439 dissolution, over 20x higher than unprocessed drug. Both the spray dried and lyophilized NP powders achieve OZ439 supersaturation during this test.

Right: Dissolution kinetics of spray dried NPs across 6 months in storage in an oven at 40 ºC and 75% relative humidity. Dissolution kinetics, along with DSC and XRD profiles, remain constant, indicating that in NPs, drug remains stable in its fast-dissolving amorphous form despite the harsh conditions.

565802 Theranostic Nanocarriers Combining High Drug Loading, Heat Release, and Magnetic Particle Imaging

Introduction: Here we report formulation and characterization of theranostic nanocarriers that combine high and tunable loading of the chemotherapeutic drug doxorubicin with iron oxide nanoparticles that enable heat release in response to alternating magnetic field and unambiguous, quantitative, tomographic imaging *in vivo* through Magnetic Particle Imaging (MPI). The nanocarriers are obtained by Flash NanoPrecipitation (FNP), which enables scalable production. The doxorubicin and iron oxide loading can be easily tuned through the inlet stream concentrations and overall particle size can be tuned through the concentration of an outer amphiphilic block copolymer corona. We demonstrate high (up to 50% w/w) doxorubicin loading, pH tunable release, significant heat release in response to applied alternating magnetic fields (AMF), and ability to image through MPI.

Materials and Methods: To facilitate doxorubicin loading through FNP, we synthesized a doxorubicin prodrug (proDox) consisting of a doxorubicin-imine conjugate. The hydrolysis of proDox to the parent drug doxorubicin was evaluated via NMR as a function of temperature and water content. Flash Nanoprecipitation (FNP) with PEG-PLA block-copolymers, iron oxide nanoparticles, and proDox was used to create the nanocarriers. Magnetic filtration was used for purification after FNP, which is rapid and scalable. Nanocarriers were prepared under a variety of FNP conditions, including variation of proDox concentration, nature of outer block copolymer corona, and presence of a hydrophobic homopolymer. The concentration of drug in the nanocarriers was evaluated spectrophotometrically. Release tests were done at 37°C in a shaking incubator with aqueous media at pH values of 4, 6, and 7.4 to represent lysosomal, intratumoral, and vasculature pH conditions, respectively. Toxicity to cancer cells was assessed by seeding MDA-MB-231 cells at 5,000 cells/well onto a 96-well plate, letting the cells grow for 48 hours, exposing cells to free doxorubicin or doxorubicin-loaded MCNCs for 48 hours, and then performing a metabolic assay to quantify relative viability. Heat release from the nanocarriers was evaluated using a D5 Series Driver (nanoscale Biomagnetics) equipped with calorimetry coil. MPI scans were done with a Momentum™ Magnetic Particle Imaging scanner (Magnetic Insight).

Results and Discussion: Drug loading was found to be highly tunable via changing the concentration of proDox in the FNP inlet streams. All 5 conditions resulted in statistically significant differences in loading, with the loading increasing as the concentration of proDox increases. The loading ranged from 5 to 50 % w/w doxorubicin. Release studies showed that the release rate is fast and highly dependent on pH. The release rate was insensitive to the nature and molecular weight of the block copolymer corona used to cap the nanocarriers. This suggests that hydrolysis of the proDox is the rate limiting step for release. The nanocarriers were toxic to cancer cells *in vitro*, with an IC50 value of 1,100 nM, compared to 200 nM for free doxorubicin. The heating rate of the nanocarriers was tunable through the amplitude and frequency of the applied AMF. MPI scans showed that the nanocarriers can be imaged, which suggests further studies of pharmacokinetics and biodistribution using MPI, which provides for unambiguous, quantitative, tomographic, and longitudinal imaging of iron oxide nanoparticle biodistribution.

Conclusions: This study shows that magnetic nanoparticles can be combined with hydrophobic prodrugs to form nanocarriers that combine high drug loading, ability to release heat in response to alternating magnetic fields, and quantitative imaging through MPI.

Acknowledgements: We acknowledge funding from NSF HRD-1345156.

577244 Translation of Nanoparticle Therapeutics from Laboratory Discovery to Clinical Scale: Scaled-up Production and Solidification

Malaria is a major cause of mortality in the developing tropical regions and its effective low-cost therapies are in much need. Lumefantrine is a Biopharmaceutics Classification System class II drug and can be used as a cure for malaria. We have developed in the past low-cost lumefantrine encapsulated nanoparticles (NPs) using stabilizers such as zein/casein and HPMCAS via Flash Nanoprecipitation (FNP) method. A continuous production line of NPs is set up to transfer the nanoparticle production from laboratory scale to clinical level. Using mixers of three different scales, the production rate of NPs is increased from a batch of a few milligrams to a continuous production of a hundred kilograms per day. An ultrafiltration system has been constructed to increase the NP concentration after FNP, which greatly improves the NP solidification efficiency. The shear rate and transmembrane pressure of NP flow within different types of filters are carefully controlled to prevent NP aggregation and filter cake accumulation. Next, scalable and continuous spray drying was applied to NP suspension of increased concentration to obtain dried powders with long-term storage stability. The dissolution kinetics of dried NP powders are showed to be significantly improved compared to that of crystalline lumefantrine in simulated fasted and fed intestine media. These results demonstrate complete translation of therapeutic nanoparticles from laboratory discovery to large scale clinical production.

575438 Engineering PLGA Nanoparticles to Encapsulate Large CRISPR-Cas9 Plasmids

Gene therapy involves the permanent alteration of DNA to fix genetic diseases. The Clustered Regularly Interspaced Short Palindromic Repeats (CRISPR) technique has received much attention recently due to its potential for revolutionizing targeted genome editing with unprecedented precision and control. However, one of the main setbacks to implementing this new process is the difficulty in effective delivery of high molecular weight DNA into the cell. This CRISPR technique typically uses a circular plasmid DNA, which is hard to transfect into the cell due to natural defense mechanisms that inhibit foreign DNA materials from entering.

Previous gene transfection approaches have used both chemical and physical stimuli, but these often destabilize cellular membranes resulting in unacceptable cytotoxicity especially in test animals.[1] We hypothesize that using polymer nanoparticles will protect the plasmid from degradation en route and increase delivery into the cell without the harmful side effects seen using other methods for transfection.

Using a modified nanoprecipitation method[2], a poly(lactic-co-glycolic acid) (PLGA) nanoparticle carrier with a diameter ~ 160 nm was engineered to deliver a CRISPR-Cas9 plasmid designed to correct a single point mutation in the *Tlr4* gene in mice that cause them to not react to endotoxins, such as lipopolysaccharides (LPS). The PLGA carrier was also fluorescently labeled by encapsulation of a commercially available fluorophore, 6,13-Bis(triisopropylsilylethynl) pentacene (TIPS pentacene), that has not been used in the biomedical field until recently.[3]

However, before even getting to animal studies, the particles must first be appropriately designed, fabricated, and characterized to ensure that the requirements of the application are met. The focus of the research to be presented is on the particle processing and engineering that precedes the cell and animal studies. This work addresses the questions: What is the best solvent system to use for particle fabrication? What are the resulting sizes and zeta potentials? What is the composition of the polymeric carrier as well as what was successfully loaded? How can the particles be processed for long-term storage stability? Once these are answered, then the method can be optimized to increase DNA loading. An amine-end capped PLGA was used to form nanoparticles loaded with 1.6 wt% DNA, corresponding to an encapsulation efficiency of 80%. This loading is comparable to the ~1 wt% loadings seen for plasmids half the size.[2] In release studies, most of the DNA was released within the first 24 hours and corresponded to ~2-3 plasmid copies released per nanoparticle. *In vitro* experiments conducted with murine bone marrow derived macrophages demonstrated that after 24 hours of incubation with the PLGA-encapsulated CRISPR plasmids, the majority of cells were positive for TIPS pentacene and the protein Cas9 was detectable within the cells. These results indicate that the process described here will be effective for future *in vivo* applications of the CRISPR-Cas9 system.

1. Bauer, M.; Kristensen, B. W.; Meyer, M.; Gasser, T.; Widmer, H. R.; Zimmer, J.; Ueffing, M., Toxic effects of lipid-mediated gene transfer in ventral mesencephalic explant cultures. *Basic Clin Pharmacol* **2006,** *98* (4), 395-400.
2. Niu, X. M.; Zou, W. W.; Liu, C. X.; Zhang, N.; Fu, C. H., Modified nanoprecipitation method to fabricate DNA-loaded PLGA nanoparticles. *Drug Dev Ind Pharm* **2009,** *35* (11), 1375-1383.
3. McDaniel, D. K.; Jo, A.; Ringel-Scaia, V. M.; Coutermarsh-Ott, S.; Rothschild, D. E.; Powell, M. D.; Zhang, R.; Long, T. E.; Oestreich, K. J.; Riffle, J. S.; Davis, R. M.; Allen, I. C., TIPS pentacene loaded PEO-PDLLA core-shell nanoparticles have similar cellular uptake dynamics in M1 and M2 macrophages and in corresponding in vivo microenvironments. *Nanomedicine: Nanotechnology, Biology and Medicine* **2017,** *13* (3), 1255-1266.

577990 Engineering Lipid Nanoparticles for Drug Delivery to Alleviate Oxidative Stress during Stem Cell Transplant Therapy

The success of stem cell therapy is dependent on the survival of stem cells against oxidative stress and inflammation. Due to shear stress during injection, reactive oxygen species (ROS) are produced resulting in apoptosis and release of inflammatory mediators. Vitamin E has been well known for its anti-oxidant properties and thus can be used as a drug in this study. In this study, we will describe the engineering of clinically translatable lipid based nanoparticles (LNPs) encapsulated with Vitamin E to protect iPS cells target against hydrogen peroxide induced oxidative stress. LNPs were characterized by both photon correlation spectroscopy to determine particle size distribution, hydrodynamic diameter, and polydispersity, and by zeta potential measurements to determine nanoparticle surface charge characteristics and to quantify colloidal stability. The encapsulation studies were performed using the standard curve for absorbance measurement of vitamin E at 295 nm wavelength and we achieved about 75 % encapsulation efficiency. The preliminary uptake studies using stem cells and FITC tagged LNPs indicated that the nanocarriers were internalized into the cytosol. The oxidative stress was induced by treating iPS cells with 100 µM H_2O_2 with or without LNP encapsulated Vitamin E. The viability of iPS cells was enhanced after LNP-vitamin E treatment. In addition, ROS production was observed after vitamin E treatment. The vitamin E loaded LNPs targeted to these cells showed a significant decrease in the reactive oxidative stress induced due to peroxide. On further performing MTT assay, it showed that cells treated with vitamin E loaded particles had significantly higher viability than control groups just treated with peroxide. We are currently measuring gene expression analysis of vitamin E-treated iPS cells associated with oxidative stress and apoptosis, including *NOS2*, *Casp3*, p53, BAX, NF-kappa B, and *HIF1a* genes. The findings of this study may help in developing effective stem cell therapy for the diseases characterized by the oxidative stress and inflammation.

578760 Photoelectrochemical Sensor

The evolution of photoelectrochemical (PEC) bioanalysis has resulted in substantial progress in its analytical performance and biodetection applications. PEC sensor represents a unique means for chemical and biological detection, with foci of optimizing semiconductor composition and electronic structures, surface functionalization layers, and chemical detection methods. Here, I have briefly discussed my recent developments of nanowire-based PEC sensing, with particular emphasis on three main detection mechanisms and corresponding examples. I have also demonstrated the use of the PEC sensing of real-time molecular reaction kinetic measurements, as well as direct interfacing of living cells and probing of cellular functions. My work will serve as a useful source to inform the interested audience of the latest developments and applications in the field of PEC bioanalysis.

580152 High-Throughput Sensing of Single-Bacterium Growth: Toward Rapid Antibiotic Susceptibility Testing

Accurately measuring the number of viable microorganisms plays an essential role in microbiological studies including antibiotic susceptibility testing (AST). Since conventional culture-based AST methods are too time-consuming (typically 24-72 h), rapid AST is urgently needed for preventing the increasing emergence and spread of antibiotic resistant infections. Although several phenotypic antibiotic resistance sensing modalities are able to reduce the AST time to a few hours or less, concerning the biological heterogeneity, their accuracy or limit of detection are hampered by low throughput. Here, we present a rapid AST method based on whole slide imaging (WSI)-enabled high-throughput sensing of single-bacterium growth, conducted on custom-built on-glass-slide culturing device. The time for determining the minimum inhibitory concentration (MIC) was theoretically shortest that ensures the growth of each individual cell present in a large population is inhibited. Reliable MIC of *E. cloacae* against gentamicin was obtained within 1 h, while the gold standard broth dilution method required at least 16 h for the same result. Owing to the feature of high-throughput quantitative analysis, our approach can be further applied to timely identifying the rising antibiotic resistance adapted to the treatment and guiding the adjustment of treatment strategy to prevent the aggravation of antibiotic resistance.

570379 Ionophore-Based Optical Sensors Incorporating Long-Lifetime Phosphorescent Microparticles for Background-Free Ion Detection in Biological Samples

Fluorescent sensors are ideal for biological applications due to their low invasiveness and miniaturizability. However, signal interference from autofluorescence and scattering limits their practical application. Long-lifetime phosphorescent materials have emission lifetimes on the order of hours or even days, and have long been used commercial in street signs and 'glow-in-the-dark' materials. More recently these materials have attracted interest as agents for biological imaging and sensing as a method for reducing background noise, since, by programming a delay between excitation and emission collection, the short-lived signals from scattering and autofluorescence can be avoided completely. In this work we demonstrate an ionophore-based optical sensor incorporating long-lifetime phosphorescent microparticles for background-free sodium detection in biological samples. We couple the phosphorescent microparticles to a pH-sensitive quencher dye and sodium-binding ionophore to create a dynamic and reversible optode film selective for sodium. We also characterize the sensor in terms of selectivity and stability and demonstrate its usefulness to detect sodium levels in human serum.

576752 Quantification of Biomarkers at Point of Need Using Metal Enhanced Fluorescence and Surface Acoustic Waves

Point of need (PON) quantification of cancer biomarkers such as those used in monitoring disease progression requires sensing at low (pg/ml) limits of detection from body fluids such as blood. For example, proteins such as CEA and CA19-9 are monitored by clinicians for progression of pancreatic cancer, and CA125 is monitored for ovarian cancer progression. Proteins such as PSA are used for initial screening for prostate cancer. All such tests are currently performed in clinical laboratories, off site. The capability to provide such monitoring at the clinician's office or hospital is challenging, as these proteins need quantification at low, pg/ml to ng/ml levels from a body fluid such as blood, which contains interfering proteins at orders of magnitude higher concentrations. We present concepts and results for such detection using plasmonic (metal) enhancement of fluorescence for lowering detection limits, and Rayleigh surface acoustic waves (SAWs) for reducing interference from non-specific binding (NSB) of other proteins in an immunofluorescence format. These same Rayleigh SAWs are utilized in our platform to reduce incubation time, by inducing sample mixing.

In metal enhanced fluorescence (MEF), fluorescence intensity is increased by metallic nanostructures in the proximity of the fluorophore, leading to lower limits of detection. Recent methods for creating metallic nanostructures, such as e-beam nanolithography, colloidal lithography, and colloidal self-assembling, require complicated processes and have various drawbacks. In this work, we have developed a simple nanostructure fabrication process combining thermal annealing and oxide deposition. A thin silver film is treated with rapid thermal annealing to generate the nanostructures on a SAW device surface. These nanostructures are then coated with a thin silica film to protect them and also to control the distance between the metallic structures and fluorophores. This process provides a solution to achieving silver nanostructures over a large area with simple processing techniques. Immunofluorescence experiments conducted using these surfaces show significant enhancement of fluorescence intensity. The effect of initial silver film, and oxide coating thicknesses on fluorescence enhancement was studied in detail. All fabricated nanostructures showed enhancement, with the silica layer producing an additional enhancement. The optimized structure produced a significant intensity enhancement of ~19X, leading to the construction of an immuno-sensor with a limit of detection of a few 10's of pg/ml. Finite difference time domain (FDTD) simulations for the enhancement factor allowed for understanding the role of the silica coating, and optimization of the immuno-sensor.

Acoustic streaming from Rayleigh SAWs is utilized to remove NSB proteins to improve sensor signal in our PON sensor platform. We show that NSB proteins which interfere with sensor signal are removed effectively, and incubation times are reduced significantly utilizing these SAWs. The sensor electronics and optics for this proposed sensor platform can be easily reduced to a small, hand-held package, which we present, with further reduction in size possible. The platform is demonstrated with quantification of a typical biomarker, CEA, from blood samples.

586770 Chemically & Electrochemically Interactive Characterization of Biomaterials *in vitro*

The ever growing demand for advanced and high performance implantable biomedical materials entails development of better interactive in vitro characterization tools to obtain a more representative prediction of their performance in vivo. Herein cases of implantable biomaterials, for temporary bone scaffolding support and coronary stents applications have been discussed.

Temporary implants pose a great challenge where the implant is expected to degrade and dissolve after a period of time at which point it is either absorbed by the body or excreted in the urine. The two specific medical applications are the temporary scaffolding for bone fixation and the coronary stents application. This is to prevent the long term complications associated with presence of a foreign material in the body, for example inflammatory reactions, thrombosis and restenosis in stent applications associated with the permanent stainless steel stents. Research in this area has revolved around magnesium in the past decade as it is safely metabolized by human body upon degradation and yet possesses adequate mechanical and load-bearing properties to support a healing bone as a scaffold or an artery from elastic recoil after balloon angioplasty. The main shortcoming however is the rapid degradation of magnesium in physiological environment that causes premature failure of the implant. In a series of studies [1–7] magnesium alloys were designed and fabricated, the underlying degradation mechanism in simulated biological environment was unraveled and surface treatments were applied to fine-tune degradation rate and biocompatibility. Scanning electrochemical microscopy was extensively used as a uniquely interactive characterization technique to examine degradation of magnesium at high spatial resolution in-situ. A new bio-sensing method was introduced for detection of H2 which enabled detection of localized degradation. Localized degradation is one of the main reasons for the premature failure of load-bearing structures. The new insight into the degradation mechanism allowed for an informed exploration of surface treatments to fine-tune degradation rate of magnesium in a physiological environment. In particular, ionic liquid surface treatment, conversion coatings based on rare earth metal and a biodegradable conducting polymer coating were investigated in order to suppress the initial rapid degradation, alkalization and H2 evolution while in the longer term creating a self-healing surface film capable of maintaining uniform degradation pattern. Results of these works introduced candidates with optimized bio-degradation rate that would minimize the risk of early stage alkalization and inflammatory reaction for the surrounding tissue as well as minimizing risk of premature failure to ensure adequate mechanical support is provided during the healing (of the bone) or remodeling (of the artery).

In the field of biomaterials sciences, although there are standards developed to examine the performance of materials in certain in vitro environments, there is still major need for a testing procedure with holistic approach to take into account most environmental parameters, e.g. load and cycles, ions/electrolyte, proteins, cell adhesion, mobility, temperature etc. Currently the in vitro test can be considered only as a preliminary screening test with limited ability to give the realistic picture, and a holistic simulation procedure has to be developed to measure performance in the biological condition. For example a hip and knee simulators with the tribological test properties of the materials together with full consideration of the corrosivity in the cell culture environment and other interacting elements. It is only with a holistic approach and using a multi-array of sensors that the full impact of an implant material on the host can be determined. Surface modifications are often performed on the biomedical implants to improve corrosion resistance, wear resistance, surface texture and biocompatibility. The interactions that take place at the atomic level between the surface of the implant, the host and the biological environment including all types of micromotions of the implants similar to what is expected in the human system should be studied thoroughly in order to advance implant materials with predictable function in the body.

References

[1] S.S. Jamali, S.E. Moulton, D.E. Tallman, M. Forsyth, J. Weber, G.G. Wallace, Applications of scanning electrochemical microscopy (SECM) for local characterization of AZ31 surface during corrosion in a buffered media, Corros. Sci. 86 (2014) 93–100.

[2] S.S. Jamali, S.E. Moulton, D.E. Tallman, M. Forsyth, J. Weber, G.G. Wallace, Evaluating the corrosion behaviour of Magnesium alloy in simulated biological fluid by using SECM to detect hydrogen evolution, Electrochim. Acta. 152 (2015) 294–301.
[3] S.S. Jamali, S.E. Moulton, D.E. Tallman, M. Forsyth, J. Weber, A.S. Mirabedini, et al., Corrosion protection afforded by praseodymium conversion film on Mg alloy AZNd in simulated biological fluid studied by scanning electrochemical microscopy, J. Electroanal. Chem. 739 (2015) 211–217.
[4] Y. Zhang, X. Liu, S.S. Jamali, B.R.W. Hinton, S.E. Moulton, G.G. Wallace, et al., The effect of treatment time on the ionic liquid surface film formation: Promising surface coating for Mg alloy AZ31, Surf. Coatings Technol. 296 (2016) 192–202.
[5] S.S. Jamali, S.E. Moulton, D.E. Tallman, Y. Zhao, J. Weber, G.G. Wallace, Self-healing characteristic of praseodymium conversion coating on AZNd Mg alloy studied by scanning electrochemical microscopy, Electrochem. Commun. 76 (2017) 6–9.
[6] D. Liu, G. Xu, S.S. Jamali, Y. Zhao, M. Chen, Fabrication of biodegradable HA/Mg-Zn-Ca composites and the impact of heterogeneous microstructure on mechanical properties, in vitro degradation and cytocompatibility, Bioelectrochemistry. 129 (2019) 106–115.
[7] S.S. Jamali, S.E. Moulton, Y. Zhao, S. Gambhir, M. Forsyth, G. Gordon, A biodegradable conducting polymer coating to mitigate early stage degradation of magnesium in simulated biological fluid : an electrochemical mechanistic study, ChemElectroChem. 6 (2019) 4893–4901.

578352 Solvent-Resistant Electrochemical Transistors As Dual Sensing Platforms in Aqueous and Organic Media

Organic electrochemical transistors (OECTs) are promising contemporary sensing platforms for high-fidelity transduction of (bio)chemical signals into electrical observables. In an OECT, such signals cause can ions to inject from an electrolyte into an organic semiconducting polymer channel, changing its bulk conductivity and the output drain current. Such devices provide multiple advantages that make them apt for bioelectronics applications, namely (1) operating voltages less than 1 V, (2) direct interfacing with biologically relevant electrolytes for *in vivo* application, and (3) printable fabrication, for flexible and stretchable form factors. Despite such advantages, OECTs have only achieved high signal transduction with a narrow subset of channel materials, many of which require cytotoxic cross-linkers to prevent dissolution in aqueous environments and/or exhibit poor processability in carrier solvents. Consequently, there is a strong need to develop biocompatible channel materials with high-processability.

In this work, I discuss the development of single-component OECTs which are insoluble to both aqueous and organic media, yet can be facilely solution-processed from commercial water-soluble carboxylated polythiophene (CPTs) precursors post-synthesis. Such OECTs demonstrate robust stability and high pseudocapacitances even after repeated 2-V width voltage sweeps, as well as high transduction (transconductances = 30 mS at V_{gate} = -0.5 V) in aqueous and nonaqueous systems. Changes in doping state and ion injection will also be probed using spectroelectrochemistry and electrochemical quartz crystal microbalance to further understand the origin of high transconductance in these materials. In conjunction with other electrochemical techniques, we should be able to elucidate detailed characteristics that dictate high mixed ionic-electronic transport in these materials.

563640 Investigation of Polymer-Plasticizer Blends for BTEX Sensing Using Acoustic Wave Devices

Volatile organic compounds (VOCs) are chemicals which evaporate easily and become gases at ambient temperature and pressure. The EPA has made the monitoring of VOCs mandatory due to their health impacts, which range from headaches and nausea to cancer. The current techniques used for VOC monitoring such as photo ionization, gas chromatography, mass spectroscopy, and e-noses sensors are expensive, time-intensive, demand rigorous sample preparation and cannot quantify the VOCs in the air. To employ the shortcomings of current sensor systems, quartz crystal microbalance (QCM) and surface acoustic wave (SAW) sensors based on polymer-plasticizer thin films have been developed. Acoustic wave devices are capable of detecting very low concentrations (ppm, ppb) of organic compounds in the environment. Imprinted polymers as sensing film have been used to detect BTEX gas analytes to a lower limit of 0.5ppm using SAW devices.

Traditionally, the sensitivity and performance of rubbery polymers towards hydrocarbon sensing are superior to glassy polymers however, they show low molecular selectivity due to their amorphous nature, availability of large free volume elements, and display of undesirable viscoelastic effects. Plasticization helps to modify the chemical and physical properties of the glassy polymer by decreasing the glass transition temperature, altering the pore dimensions and increasing the free volume of the polymer film thus enabling higher diffusion and sorption of the analyte molecules. This work shows that the selectivity and sensitivity of glassy polymers can be tailored by introducing a plasticizer. Poly (ethyl methacrylate) (PEMA), Poly (methyl methacrylate) (PMMA), PEMA-PMMA, PMMA-Polystyrene copolymers were modified by introducing plasticizers, diisononyl cyclohexane-1, 2-dicarboxylate (DINCH), diisooctyl azelate (DIOA), dibutyl sebacate (DBS), n-butyl stearate (BS), and di-n-butyl phthalate (DBP) to enhance BTEX sensitivity and selectivity in presence of interferants.

The objective of this study is to investigate the effect of various plasticizers on glassy polymers sensitivity by unraveling the polymer-plasticizer transient response mechanism and identifying the accurate plasticizer for superior BTEX sensing. This work demonstrates that the incorporation of suitable plasticizers into a polymer-based sensor can vastly improve the selectivity for BTEX compounds based on properties such as sorption (via thermodynamic partition coefficients), saturation dynamics, stability, and viscoelasticity of the film. A mathematical model will be incorporated to select the optimum polymer-plasticizer blend for BTEX sensing in gas and liquid phase using QCM and SAW devices.

575123 Microfabricated Silion Nanowire Array Platform for Surface Enhanced Raman Spectroscopy

This work presents a microfabricated silicon nanowire array chip for Surface Enhanced Raman Spectroscopy (SERS) to rapidly detect trace target analytes in biofluids such as serum, urine and exhaled breath. Silicon nanowire array are created by wet etching and coated with silver nanoparticle. Silver nanoparticle coated silicon nanowire array chips are characterized for surface enhanced Raman spectroscopy (SERS). The silicon nanowire array can result in significant SERS for target analyst. The chips are used to detect trace THC in exhaled breath by SERS. The results show that silicon nanowire array coated with sputtered silver nanoparticles can achieve a limit of detection of 3.1 pg of THC. The linear relationship between SERS signal and the amount of THC indicate that the method is suitable for quantification of THC concentration in breath.

572677 Invited Talk: Resolving Carbon Nanotube Structures By DNA

Many fundamental studies and technology development rely on structurally well-defined CNTs as well as corresponding coating layers. DNA-CNT library has been used to select structurally well-defined hybrids using separation method. Existing selection methods have various drawbacks. In this talk, I will present our recent work in developing new spectroscopy and separation method to obtain DNA sequences that are capable of resolving different CNT structures. I will also present a multiplex sensing scheme we call "molecular perceptron", which is based on an array of DNA-CNT hybrids to mimic human perception systems for biosensing applications.

564140 Thermomechanical Behavior of Nanodiamond Superstructures in Interlayer-Bonded Twisted Bilayer Graphene

Graphene derivatives and metamaterials, fabricated through chemical functionalization and defect engineering of graphene sheets, have major potential for technological applications because of their exceptional thermomechanical properties that can be tuned by tailoring of their nanostructural features. Here, we report a systematic computational study on the thermomechanical behavior of a class of two-dimensional (2D) carbon-based nanostructures, namely, nanodiamond superstructures in interlayer-bonded twisted bilayer graphene, formed through hydrogenation-induced interlayer covalent bonding of twisted bilayer graphene with commensurate bilayers. We have conducted a comprehensive classification of all such superstructures that can be formed with respect to the corresponding structural parameters, including the commensurate bilayer's twist angle, the interlayer bond pattern and density, and the concentration of sp^3-bonded C atoms in these superstructures. We report results for the mechanical behavior of such carbon nanocomposite superstructures based on molecular-dynamics (MD) simulations of uniaxial straining tests according to a reliable interatomic bond-order potential and establish the dependence of the superstructures' mechanical properties on the concentration of sp^3-bonded C atoms. We also define a ductility metric and demonstrate that a brittle-to-ductile transition occurs in these superstructures with increasing the concentration of sp^3-bonded C atoms beyond a critical level. The underlying ductile fracture mechanism, mediated by void formation, growth, and coalescence, in contrast to the typical brittle fracture of graphene is characterized, and the superior mechanical response to uniaxial straining of the ductile nanodiamond superstructures is demonstrated.

Furthermore, we study the mechanical response of these superstructures to nanoidentation based on MD simulations of nanoindentation tests. We establish the dependences of the elastic modulus and the hardness of the superstructures on their structural parameters as well as on indentation parameters. The resulting structural response of the superstructures under nanoindentation testing also is characterized in detail. Finally, we report results for the lattice thermal conductivity of these superstructures based on non-equilibrium MD simulations of thermal transport. We report a significant reduction in the lattice thermal conductivity with increasing concentration of sp^3-bonded C atoms in the superstructures, which is promising for thermal management applications, and establish the dependence of the lattice thermal conductivity on the full range of structural parameters for these interlayer-bonded bilayer materials.

562664 Multiwalled Carbon Nanotube Functionalization and Characterization: Evidence for Lactone and Ester Formation

The presence and density of surface functions on carbon materials greatly affect their utility in adsorption processes (e.g., for heavy metals or organics), in catalysis (e.g., as supports or in metal-free catalysis), and as carbon-fiber composites (e.g., increasing mechanical strength). Oxygen-based surface functions are the most commonly studied and introduced through treatment in strong oxidizing agents or exposure to plasma, which are frequently used to functionalize carbon materials. Treatment with plasma is a quick and environmentally-benign process for introducing oxygen-based surface functions on carbon materials depending on the source composition; however, the precise identity of the surface functions that form is still unclear despite significant efforts in this area. The most commonly implicated moieties on carbon surfaces are ethers (C-O-C), alcohols (-OH), carboxylic acids (-COOH), and esters or lactones (O=C-O). Commonly implemented spectroscopic techniques (e.g., NMR, XPS, IR) are limited due to weak signal from low density of functions or lack chemical specificity and resolution to distinguish among plausible functions. Here, we present a reactive-titration method for identifying and quantifying the density of -COOH and O=C-O surface functions on O_2 plasma treated multiwalled carbon nanotubes (MWCNT), which provides a methodology for characterizing these materials that will allow researchers to better-understand how the presence, proximity, and density of these functions can impact the physicochemical properties of a material.

The surface of MWCNTs were functionalized by treatment with O_2 plasma in a capacitively-coupled plasma chamber (300 W, 220 mTorr, 25 cm^{-3} min^{-1}, 0 – 4 min; March CS-1701, Nordson March). The elemental abundance of oxygen was evaluated by X-ray photoelectron spectroscopy, which shows that the oxygen content of MWCNT increases after plasma treatment (oxygen peak fraction from 17% to 24 %). Peak fitting of C 1s features in the XP spectra show that the abundance of C-O and O=C-O increase slightly (5.5% to 6.3% and 4.3% to 5.0%, respectively), while the density of C=O moieties decreases (3.3% to 2.6%) after plasma treatment (4 min). Unfortunately, the identity of the chemical functions that contribute to these XP features cannot be assigned clearly by XPS alone. Therefore, we have developed potentiometric titration protocols for doing so.

Chemical reactions induced via titrations (e.g., protonation, hydrolysis) with sodium hydroxide can yield important information regarding the nature of the functional group (e.g., pKa).[1-2] Potentiometric titrations (0.01 M NaOH as titrant) of plasma-treated MWCNT (pt-MWCNT) show a single gaussian feature for acidic function (0.54 mmol g^{-1}) with a pKa of 6.3, while only small amount of Bronsted acid sites (0.07 mmol g^{-1}) are observed for untreated MWCNT between 3 – 11 pH units. Upon re-acidification (0.01 M HCl) and sequential titration with NaOH, the acidic function detected on pt-MWCNT does not reappear during titration, which suggests that this surface function irreversibly reacted with NaOH and is not a -COOH moiety. More likely, this feature is related to a lactone or ester group that undergoes saponification upon titration with NaOH. The pKa of common -COOH groups (3 – 6) should be observable after acidification in sequential titration, which suggests that the partially oxidized MWCNT surface acts as an electron-withdrawing functional group vicinal to the -COOH and gives a lower pKa for the -COOH (and one below the pH titration window in water, i.e., 3 – 11). To test this hypothesis, we used ethyl pyruvate (an α-ketoester) as a probe molecule, which showed a single feature at pKa of 9.2 upon initial titration, with no features observed in repeated titrations. The same trend of ethyl pyruvate and pt-MWCNT in titration suggests that treatment of MWCNT in O_2 plasma may produce α-ketoester surface functions, which can undergo saponification to produce alcohols and α-ketoacids out of pH window of potentiometric titration. We conclude that oxygen plasma treatments introduce lactone or ester functions onto MWCNT, and their close vicinity to electron withdrawn groups decreases the pKa of -COOH that form by hydrolysis.

1. Contescu, C.; Jagiello, J.; Schwarz, J. A., Heterogeneity of proton binding sites at the oxide/solution interface. *Langmuir* **1993,** *9* (7), 1754-1765.
2. Contescu, A.; Contescu, C.; Putyera, K.; Schwarz, J. A., Surface acidity of carbons characterized by

their continuous pK distribution and Boehm titration. *Carbon* **1997,** *35* (1), 83-94.

576076 Growth and Modification of Carbon Nanotubes By Atomic Layer Deposition of Ni and Cu for Spaceflight Instruments and Terrestrial Applications

Growth and Modification of Carbon Nanotubes by Atomic Layer Deposition of Ni and Cu for Spaceflight Instruments and Terrestrial Applications

Vivek Dwivedi[1], Henry Degroh[2], Raymond A. Adomaitis[3], Hossein Salami[3], Alan Uy[3], Aarathi Vadapalli[3], Brianna Sheard[3]

1 NASA Goddard Spaceflight Center, USA

2 NASA Glenn Research Center, USA

3 University of Maryland, USA

ALD is a cost-effective nanoadditive-manufacturing technique that allows for the conformal coating of substrates with atomic control in a benign temperature and pressure environment. Through the introduction of paired precursor gases, thin films can be deposited on a myriad of substrates regardless of topography. By providing atomic layer control, where single layers of atoms can be deposited, the fabrication of metal transparent films, precise nano-laminates, and coatings of nano-channels and pores is achievable. A characteristic of the surface adsorption and reaction mechanisms in ALD is that they are normally self-limiting, allowing for atomically accurate control of nanometer (nm) thickness. Therefore, high uniformity and precise thickness control make ALD an attractive process for the creation of novel optical and other nano-scale devices. The applicability of ALD as a mode for depositing the catalytic layer for the growth of carbon nanotubes, CNTs, and using ALD to directly modify a CNT substrate are discussed.

Stray light optics for space platform instruments require components of the optical bench to be coated with a material that suppresses light reflectance allowing for the imaging of extremely dimmed objects. Coronagraphs are key heliophysics instruments because they image coronal mass ejections (CMEs), which are the most energetic phenomena on the Sun. CMEs have wide-ranging impact on the heliosphere, from interplanetary spacecraft to Earth-orbiting satellites, communications, and astronaut safety—in short, they are major drivers of space weather. Coronagraphs reveal the structure and speed of CMEs when they are near the Sun and provide essential input to predictive space weather models. In a typical space-based coronagraph, an external occulter blocks light from the disk of the Sun so that the corona (about a million times dimmer) can be imaged. The occulter must suppress both diffracted light and stray light. Diffraction is controlled by careful design of the occulter geometry, typically a curved surface and the stray light reflectance is controlled by the occulter coating, typically anodized black. A promising advance in stray light control is the application of carbon nanotubes. Carbon nanotubes have shown to be 10 to 100 times blacker than anodized black from the Near UV to Far Infrared. In this presentation we will show the utilization of ALD to deposit a catalytic layer of Ni for the growth of CNTs for substrates with varying curvature with enhanced reflectance properties.

There is great need for the increase in electrical conductivity of wire, and also for the decrease in wire density. Increases in electrical conductivity are needed to improve the function of electronics, electric motors and everything else that runs on electricity, as well as cables that transmit electricity. An increase in conductivity results in greater efficiency due to lower power losses. The state-of-the art in wire has not change much over the last 100 years. The conductor material of all applicable wire and cable is made of either copper (Cu) or aluminum (Al). Copper has superior conductivity and is used when space is limited, like in the windings of an electric motor. It is believed that contributions to the state-of-the-art can be made through improvements in conductivity and/or density for Cu. This improvement can be accomplished by adding something to Cu that is more conductive and/or less dense, and at the same time, not ruin the conductivity of the Cu. The only material known to man that has significantly higher conductivity and lower density than Cu is ballistically conducting, metallic carbon nanotubes (CNT). ALD of conformal metallic copper films is an important research area where such films are needed in the miniaturization and portability of electronic devices. However, the miniaturization from the micro to nano regime has led to Joule heating becoming a major concern in interconnect applications. The conformal deposition on most high aspect ratio substrate geometries is a challenge that ALD can solve.

577085 Growth of Small Diameter Carbon Nanotubes Using High-Melting Point Transition Metal Catalyst Promoters

Growth of Small Diameter Carbon Nanotubes Using High-Melting Point Transition Metal Catalyst Promoters

Brian Everhart[1], Rahul Rao[2], Benji Maruyama[2], Placidus Amama[1]

[1]*Tim Taylor Department of Chemical Engineering, Kansas State University,*

Manhattan, KS, 66506

[2]*Air Force Research Laboratory, Wright-Patterson Air Force Base, OH, 45433*

Synthesis of small diameter single wall carbon nanotubes (SWNTs) has long been a goal within the fields of catalysis and nanotechnology. Current as-grown SWNTs vary in diameter and chirality, resulting in variations in structural and electronic properties. Typically, these tubes are comprised of approximately two-thirds semiconducting tubes and one-third metallic tubes. Furthermore, larger diameter semiconducting SWNTs (s-SWNTs) have very small bandgaps, as the bandgap is inversely proportional to diameter. Numerous applications, including photovoltaics and photocatalysis, require s-SWNTs with relatively large bandgaps, necessitating the growth of small diameter tubes. Commercially available s-SWNTs are incredibly expensive, as they require extensive post-growth processing. Other methods of achieving high selectivities of s-SWNTs and m-SWNTs involve controlling tube diameter via supported catalyst on substrates. By using high melting-point transition metals, such as ruthenium, as catalyst promoters, small diameter tubes, including high band-gap s-SWNTs can be obtained. Experiments have been performed in the Autonomous Research System (ARES), allowing for high throughput experimentation to probe the multi-parameter reaction space that's intrinsic to CNT growth. Ruthenium has been used as a growth promoter on cobalt catalyst to show substantial reduction in tube diameters, confirmed by multi-excitation Raman spectroscopy. Increasing the amount of ruthenium beyond a certain threshold, however, negatively impacts CNT growth.

577814 3D Layer-By-Layer Electrostatic Assembly of Graphene-Noble Metal Thin Films for Energy Storage and Conversion Applications

3D Layer-by-Layer Electrostatic Assembly of Graphene-Noble Metal Thin Films for Energy Storage and Conversion Applications

Mark H. Jaskot, Joshua M. Musiol, Kamil Woronowicz, Pamela L. Sheehan[1], Preston C. Haney[1], Harry L. Moore[1], F. John Burpo, Enoch A. Nagelli*

Department of Chemistry & Life Science, Chemical Engineering Program

United States Military Academy, West Point, NY 10996

[1]Explosive Ordnance Disposal, Demilitarization & Experimental Directorate, Army Futures

Command, RDECOM-ARDEC, Picatinny Arsenal, NJ

*Corresponding PI: Dr. Enoch Nagelli, **Email:** enoch.nagelli@westpoint.edu

Carbon nanomaterials and noble metal nanoparticles constitute an ideal nanostructured composite platform for integration with other solid-state electrodes for energy storage and conversion applications. Self-assembly has been recognized as an effective strategy for the bottom–up synthesis of 3D macrostructures using graphene and CNTs as building blocks. However, all solution-based assembly techniques as a scalable platform for 3D thin film nanomaterials design have received limited attention. Here we propose a novel, simplistic, and scalable methodology utilizing a solution based electrostatic self-assembly technique for the formation of novel 3D noble metal nanoparticle composites with both graphene and carbon nanotubes. For example, graphene oxide and noble metal thin film nanocomposites were synthesized using electrostatic coordination between the carboxylic acid edge functional groups and platinum (Pt^{2+}) or palladium (Pd^{2+}) ions via substrate solution casting and spin coating. The chemical reduction of the resulting nanocomposite films formed transparent, conductive, and free-standing thin films comprised of interconnected graphene and platinum or palladium nanoparticles, held together with poly(methyl methacrylate) (PMMA). This solution-based platform materials design method allows for the controlled functionalization and assembly of any noble metal nanoparticles onto 2D graphene sheets or 1D CNTs while maintaining its 3D nanostructure with direction-specific and location-specific catalytic properties. In addition, this platform thin film approach can lead to the tunable design of multi-dimensional pillared nanocomposites via layer-by-layer (LBL) stacking of dissimilar PMMA-supported carbon nanomaterials functionalized with various noble metal nanoparticles demonstrating controlled macro-assembly of nanostructured films. This method of assembly from site-specific noble metal and carbon nanomaterial integration into multi-dimensional nanocomposite electrode thin films demonstrate the macro-scale control needed for the development of next generation miniaturized energy storage and conversion devices.

566951 Graphene Oxide-Polysulfone Mixed Matrix Membranes: Effect of Graphene Functionalization and Loading on Oil-Water Separation Performance and Membrane Fouling

Microfiltration and ultrafiltration polymeric membranes plays significant role in water treatment and water desalination pre-treatment applications. The performance of these membranes is characterized by their flux, percentage rejection of impurities and resistance to inorganic, organic, and biofouling. In this presentation, the performance of polysulfone (PS) ultrafiltration (UF) membranes is enhanced by incorporating small fractions (< 0.8 wt.%) of graphene oxide functionalized with amine (a-GO) and carboxylic (cGO) groups. Pure PS and PS mixed matrix membranes (MMM) containing a-GO and c-GO are fabricated by the phase inversion method and membrane hydrophilicity, morphology, porosity, and mechanical properties are investigated using contact angle measurements, SEM, AFM, and dynamic mechanical analysis. The impacts of GO functionalization on the separation of oil-water emulsion and protein (bovine serum albumin (BSA)) fouling were investigated. The hydrophilicity of the membrane increased with increasing the concentration of the functionalized GO, regardless of the functionalization type. On the other hand, the flux and rejection was increased with GO loading up to an optimum loading of ~ 0.2 wt.% and then declined with higher concentrations. Moreover, the addition of f-GO leads to not only enhanced permeability and rejection but also significant improvement in the mechanical properties and durability compared to pristine PS membrane. The antifouling testing of the fabricated membranes was assessed by the filtration of bovine serum albumin (BSA) solution and the f-GO membranes demonstrated better fouling resistances compared to pristine PS membrane. Our results indicates that f-GO membranes have outstanding potential in the field of water treatment as they offer significant enhancement in the performance, mechanical properties and antifouling characteristics at very low concentration that does not affect the theological properties of the PS-f-GO formulation and therefore requires no alteration to the current commercial fabrication process.

564153 Mechanical Behavior of Graphene Nanomeshes

Graphene-based nanomaterials have exceptional electronic, mechanical, and thermal properties that can be tuned by tailoring their nanostructural features. Graphene nanomeshes (GNMs) are ordered, defect-engineered graphene nanostructures consisting of periodic arrays of nanopores in the graphene lattice with neck widths less than 10 nm. The electronic, transport, and mechanical properties of GNMs can be tuned by varying the structural, chemical, and architectural parameters of the nanomeshes, such as their porosity, pore lattice structure, pore morphology, as well as the type and extent of their pore edge termination.

Here, we report the results of a comprehensive study of the mechanical response of GNMs to uniaxial tensile straining and determine their mechanical properties based on molecular-dynamics (MD) simulations of dynamic deformation tests according to a reliable bond-order interatomic potential. We establish the dependences of the GNMs' ultimate tensile strength, fracture strain, and fracture toughness on the nanomesh porosity and derive scaling laws for the strength-density relation of the GNMs. We place special emphasis on how the above properties are affected by the GNMs' pore morphology and pore edge termination with H atoms. The underlying mechanisms of crack initiation and propagation, and of nanomesh failure are characterized in detail.

We also study the mechanical and structural response of GNMs to nanoindentation based on MD simulations of nanoindentation tests. We demonstrate that the elastic modulus and the hardness of the GNMs decrease monotonically with increasing nanomesh porosity and derive modulus-density and hardness-density scaling laws. We also examine the effects on the hardness and stiffness of the GNMs of the termination of the GNMs' pore edges by hydrogenation. The underlying mechanisms of nanomesh structural response upon nanoindentation also are characterized in detail over the full range of the GNMs' structural parameters examined.

577442 Continuous Green Millifluidic Synthesis of Ag Nanostructures

The synthesis of nanomaterials can be done by several chemical and physical methods including but not limited to microwave assisted, hydrothermal, heat evaporation, chemical reduction, photochemical reduction, electrochemical reduction, and so on. While nanomaterials can be successfully synthesized using these techniques, there are numerous drawbacks including high cost and the application of toxic solvents and external reducing agents, excessive material conversion, sophisticated equipment, and high-energy requirements. However, green synthesis approaches aim to achieve sustainable development of nanomaterials through the application of natural products such as reducing sugars with inherent growth terminating, stabilizing, and capping characteristics. Among different metal nanoparticles, silver nanoparticles (AgNPs) have unique and outstanding shape and size-dependent electrical, thermal, optical, catalytic, optoelectronic, and biological characteristics, which make them exceptional for several applications in a variety of fields. The morphology and structure characteristics of Ag nanostructure including size, uniformity, aspect ratio, and purity have a strong effect on the final performance of Ag nanostructure-based devices. However, green synthesis control in producing Ag nanostructure is still a challenge in batch reactor along with batch-to-batch variability. We have shown that millifluidic reactor can overcome many of the challenges in batch synthesis through control of a uniform chemical and thermal environment in a small reaction volume with the goal of reducing their final cost by an order of magnitude through the application of less expensive water based reducing agents at lower reaction temperature in a continuous manner.

The purpose of this work is continuous green synthesis of Ag nanostructures using sugar and different artificial sweeteners as reducing agent to maximize yield of synthesized AgNPs and control their morphology. Aqueous solution of $AgNO_3$ has been used as Ag precursor in the presence of white sugar, brown sugar, and several artificial sweeteners such as Splenda, Sweet'n Low, Caribou Coffee, and Whole Earth as reducing agent to synthesize AgNPs at low temperature in a millifluidic reactor. Transmission Electron Microscopy (TEM), High Resolution Transmission Electron Microscopy (HRTEM), Energy-Dispersive X-ray Spectroscopy (EDX), and Scanning Electron Microscopy (SEM) have been utilized to show the morphology and size of the synthesized AgNPs. It was shown that the Maui Raw Turbinado brown sugar can act as both a reducing agent and a capping agent to synthesize One-dimensional (1D) Ag nanostructures even in the absence of a capping agent such as polyvinyl pyrrolidine (PVP). The main ingredients of this reducing agent have been used in a small-scale batch reactor for reaction mechanism investigation by using TEM characterization and *in-situ* Ultraviolet–visible (UV-vis) spectroscopy to find the functional groups that allow the production of 1D Ag nanostructures using this specific brown sugar.

569573 Green Synthesis of Palladium Nanomaterials Via Biological Templates Using Tmv and Bsmv Virus-like Particles with Enhanced Stability and Customized Dimensions

In recent years, there has been increased interest in the use of naturally occurring biomolecules as biotemplates in the fabrication of nanomaterials. The use of the self-assembly biological structures forms one-dimensional metal nanorods and nanowires with monodisperse structures. The virus-like particles (VLPs), which are the self-assembly of virus particles, of Tobacco Mosaic Virus (TMV) and Barley Stripe Mosaic Virus (BSMV) serve as biotemplates to understand the direct synthesis of palladium nanomaterials. The TMV, BSMV VLPs and their engineered variants are produced from bacterial *Escherichia Coli* system. Our recent work has focused on the stability control and the length customization of BSMV VLPs in order to expand the diverse industrial applications, such as sensing, catalysis, electronics and medical therapy. The optimization of VLP protein extraction and purification process are investigated to be compared to plant virus particles. More uniform palladium nanorods with controllable shapes are synthesized via biomineralization using hydrothermal process in the absence of external reducing agents. The VLPs are characterized with transmission electron spectroscopy, X-ray absorption spectroscopy and energy dispersive X-ray spectrometry to provide the insights into the mineralization process. Briefly, biotemplating using BSMV VLPs expressed from *E. Coli.* which has the genetic modification malleability, various lengths and enhanced stability to generate controllable monodisperse metal nanomaterials is promising in the bottom-up nanotechnology.

575934 Scalable Biomineralization of CdS Quantum Dots By Immobilized Cystathionine γ-Lyase

The typical high-temperature, organic phase synthesis of quantum dots (QDs) leads to respective high energy demand and expensive and toxic chemical usage that limit the process scalability required to meet the production needs for promising commercial QD applications.[1–3] Biomineralization-based routes to size-controlled QDs offer an alternative, 'green', and potentially scalable synthesis strategy, characterized by low temperatures and aqueous phase processing. Biomineralization of size-tunable CdS QDs, for example, is possible by the aqueous-phase, low-temperature enzymatic (*cystathionine gamma lyase*) turnover of the amino acid L-Cysteine to H_2S in buffered solutions of cadmium acetate.[4] Leveraging this and similar processes directly for scalable QD synthesis, however, is limited by the lack of cost-effective or feasible strategies for enzyme recovery and reuse.

Enzyme immobilization on easily recoverable substrates offers a strategy for their facile separation from product solutions and, thereby, enzyme reuse. In this study, we demonstrate the efficacy of Pyridoxal phosphate (PLP)-dependent cystathionine γ-lyase (CSE), immobilized on nanoparticulate TiO_2 substrates, for the cyclic synthesis of CdS QDs via a single-enzyme biomineralization approach. Specifically, we exploit the strong physical adsorption of CSE on nanoparticulate TiO_2 aggregates for facile CSE immobilization. This allows for cyclic enzyme recovery from reaction solutions and resuspension in fresh reactant. The immobilized CSE remains active for L-cysteine turnover to H_2S in buffered solutions of Cd-acetate, albeit at an expectedly reduced level relative to free enzyme. Following a 2 hr induction, related to finite L-cysteine adsorption on TiO_2, the immobilized CSE retains *ca.* 55% of its activity over 6 cycles. We also demonstrate how the activity of the immobilized CSE can be sustained or regenerated by simply dosing PLP into the fresh reaction solution. For example, PLP addition at the start of each cycle increases nominal specific activity of the immobilized CSE, and enables retention of *ca.* 86% of the post-induction activity over 6 cycles by mitigating substrate inhibition. While the higher enzyme activity in the case of PLP dosing leads to slightly larger CdS quantum dots relative to cycling in the absence of added PLP, the CdS QDs in each case display a remarkably consistent particle size and degree of monodispersity across all post-induction cycles. This facile strategy for immobilized enzyme-mediated QD synthesis opens exciting possibilities for the scalable, 'green', biosynthesis of consistently sized QDs.

References

1. Hu, M. Z. & Zhu, T. Semiconductor Nanocrystal Quantum Dot Synthesis Approaches Towards Large-Scale Industrial Production for Energy Applications. *Nanoscale Res. Lett.* **10,** 469 (2015).
2. Lim, H. *et al.* Continuous Purification of Colloidal Quantum Dots in Large-Scale Using Porous Electrodes in Flow Channel. *Sci. Rep.* **7,** 43581 (2017).
3. Yang, H.-J. & Tuan, H.-Y. Efficient and scalable synthesis of quantum dots using hexane as the solvent in a non-microfluidic flow reactor system. *RSC Adv.* **4,** 51926–51934 (2014).
4. Spangler, L. C., Cline, J. P., Kiely, C. J. & McIntosh, S. Low temperature aqueous synthesis of size-controlled nanocrystals through size focusing: a quantum dot biomineralization case study. *Nanoscale* **10,** 20785–20795 (2018).

568753 Nanoporous Aluminum Oxide Templates of Arbitrary Thickness on Silicon Carrier Wafer

Porous templates of anodized aluminum oxide (AAO) are of interest in terms of photonics, energy storage, and other various nanoscience applications. There is a great interest to grow AAO templates directly on silicon as a carrier wafer, however most attempts rely on film deposition techniques that are generally suited for depositions on the order of 1 um. Thicker aluminum films are desirable in order to achieve better ordering of the AAO pores through two-step anodization and for greater aspect ratio for applications where surface area is the dominant factor such as solar cell and supercapacitor devices. Thick metal films are commonly applied through electrodeposition, however this technique is not suitable for aluminum owing to its high reduction potential above that of water. Aluminum and silicon are known to alloy up to approximately 1 atomic percent silicon in aluminum. Here, AlSi alloys are used as adhesion layers between aluminum foils and silicon wafers as a solution to the thin film problem. High purity aluminum foils and silicon wafers are pressed together as a diffusion couple. The couple is then annealed under argon atmosphere at temperatures from 550°C to 650°C (from below the AlSi eutectic temperature of 577°C to just below the Al melting temperature of 660°C) for various times. Cross sections of these couples are examined under scanning electron microscopy (SEM) to determine the thickness of the diffusion region and to check for the presence of voids. The chemical composition of the diffusion region is investigated through electron probe micro-analysis. Bonded Al-AlSi-Si composites are anodized in either phosphoric or sulfuric acids using a two-step process, and the final AAO morphology is examined under SEM.

576885 Large-Scale Fabrication of 50nm Nanopore Array in Teflon Using Reactive Ion Etching Technique with Anodized Aluminum Oxide Templates As Etch Masks

Advanced cryogenic systems rely on efficient cooling mechanisms for achieving higher power density and system efficiency. Improving the convective liquid-vapor phase change heat transfer mechanism in cryogenics could lead to substantial energy savings and reduced greenhouse gas emissions. Nanoporous surface textures are known to enhance heat transfer efficiency in the boiling and quenching regimes. Also, studies have reported that Teflon coated surfaces with non-wetting cavities have improved pool boiling heat transfer efficiency. Fabricating Teflon nanopore array surfaces can potentially combine the benefits of both non-wetting and nanotexture surface heat transfer enhancements. Here, we fabricated large-scale Teflon Nanopore array surfaces using highly ordered anodized aluminum oxide (AAO) template as a plasma etch mask. The AAO templates with an average pore diameter of 50nm are fabricated by a two-step anodization process in 0.3M oxalic acid solution. Inductively coupled plasma reactive ion etching technique (ICP-RIE) is used to anisotropically etch the unmasked regions to replicate the nanopore array pattern of AAO on Teflon. In traditional AAO pattern transfer techniques, the AAO barrier layer is removed using dilute phosphoric acid solution and, the polystyrene (spin coated over AAO as a template transfer layer for additional mechanical support) is removed using a single step oxygen plasma etching. Even though straightforward, these techniques are limited to smaller areas of pattern transfer due to incomplete barrier layer pore opening across the template and presence of residual polystyrene inside the pores. In this study, we identified factors that hinder the scale-up process and, we suggested alternative novel techniques to improve the pattern transfer efficiency. It's imperative to fabricate the Teflon nanopore array over a large surface area to make them commercially viable. By following the alternative novel AAO pattern transfer techniques in this work, we were able to achieve pattern transfer over 90% of the Teflon area that was under the AAO template.

562739 Jet-Mixing Microreactor for Metal and Metal-Oxide Nanomaterial Synthesis and Scale-up Considerations

The market for nanomaterials is expected to double in the next three years, calling for the creation of innovative technologies that can produce bulk material while maintaining the quality obtained at lab scale. The properties of nanomaterials are highly dependent on the mixing dynamics of their synthesis process. Macromixing in conventional batch nanomaterial syntheses increases the overall mixing time, leading to a variability within batches and a wide particle size distribution. These problems can be circumvented by using microreactors that provide a small mixing time because of their associated micromixing and mesomixing. We have developed a simple, inexpensive continuous jet-mixing reactor that is successfully used for silver nanoparticle (Ag NP) synthesis. Ag NPs are synthesized by the common route of silver nitrate reduction with sodium borohydride, using trisodium citrate as the capping agent. It is shown that jet mixing results in a narrow particle size distribution as compared to batch. We also find that the capping agent concentration required for stable Ag NP synthesis using jet-mixing is a factor of four lower than that using batch. Ag NPs synthesized using jet-mixing have greater consistency in optical properties and a longer shelf life when monitored over one month. With the aim of scaling up future nanomaterial syntheses, fundamental studies are underway to characterize the mixing of this reactor.

575940 Synthesis of Nanostructured Cobalt Oxide Rods from Waste Lithium-Ion Batteries

Cobalt oxide (Co_3O_4) is an important p-type semiconductor with spinel structure that has applications in various fields such as catalysis, solid-state electronics, and electrochemical devices. Porous Co_3O_4 nanostructures are used as one of the cathode material components in lithium-ion batteries (LiBs). These 1-D cathode structures provide a higher surface area and shorter path length for lithium ions which leads to better electrochemical performance in LiBs. In this work, a novel synthesis process has been developed for producing porous Co_3O_4 nanorods from waste LiB cathodes. The recycling process involves metal separation and recovery from lithium cobalt oxide cathodes using oxalic acid. This process directly separates lithium and cobalt by precipitating cobalt in the form of cobalt oxalate and lithium remains in solution. Addition of hydrogen peroxide as a reducing agent during metal recovery improves metal extraction efficiency and also leads to the formation of rod-shaped cobalt oxalate nanostructures. These cobalt oxalate rods can be converted to porous rods by thermal treatment under appropriate conditions. This presentation will describe the experimental and mechanistic details of the cobalt oxide nanorod synthesis process.

571227 Continuous-Flow Microwave Synthesis of Titania Nanotubes for Moisture Harvesting

Titania nanotubes are potential materials in multiple applications due to their high specific surface area, excellent photocatalytic activity, ion-exchangeable ability and superhydrophilic property. Synthetic methods of titania nanotubes have been extensively investigated, which include assisted-template method, sol-gel process, electrochemical oxidation, hydrothermal method and microwave-assisted synthesis. In this study, we report a highly efficient method for continuous synthesis of titania nanotubes. Titania nanotubes were synthesized by using anatase TiO_2 as precursor in concentrated NaOH at different heating temperature of 100 °C ~ 130 °C, under different pressure of 2 ~ 4 bar, and for different residence time of 15 min and 30 min via microwave-assisted flow synthesis method. The as-synthesized materials were characterized by transmission electron microscopy (TEM), X-ray diffraction (XRD) and micro-Raman. The optimum synthesis temperature, pressure and residence time were found to be 120 °C, 4 bar and 15 min, respectively.

Further, the titania nanotubes were modified to provide moisture harvesting property. The moisture harvesting materials were obtained by nano-confinement of low water activity (LWA) materials within the interlayer spaces of superhydrophilic titania nanotubes. The LWA material, consisting of cyclodextrin molecules covalently linked to the titania surface, attracts water vapor from atmosphere, and then the water accumulates on the superhydrophilic titania surface draining to form droplets on the tip of the nanotubes. The modified titania nanotubes could be a promising resource to supplement the ever-increasing water need.

576983 A Starch-Mediated Synthesis of Bimetallic Silver/Gold Nanoparticles with Different Chemical Composition As Antimicrobial and Anticancer Agents

Nanomedicine is a branch of nanotechnology that is concerned with the synthesis, characterization and application of nanomaterials in the field of medicine. In recent years, the application of noble metal nanoparticles (NPs) in nanomedicine has received considerable attention due to their promising use as anticancer and antibacterial treatment agents.

The synthesis of noble metal NP systems has been reported through a variety of physical, chemical and biological methods. Out of these approaches, the wet chemical reduction method is one of the most applied, due to precise compositional, size and crystallographic control of the resulting nanomaterials. In general, this method consists of a metal salt precursor, reducing and stabilizing agents and a dispersion medium. Some examples of common chemical compounds include N, N- dimethylformamide (DMF) and sodium borohydride (NaBH4) as reducing agents, cetyltrimethylammonium bromide (CTAB) as capping agents, and toluene or chloroform as solvent. Although the use of these compounds often yields better control of the final product, they may leave toxic contaminants. For this reason, their use in medical applications is potentially dangerous due to toxic effects on the environment and biological systems. An alternative to the usage of toxic reagents is the employ of plant extracts, unfortunately, this route requires a complex purification process and, in many cases, there is a necessity to use toxic solvents in the determination of the active molecules that play the role as reducing and capping agents of the NPs.

Therefore, a simpler eco-friendly method has been proposed for the synthesis of colloidal noble metal NPs, in which biocompatible starch acts as both the capping and reducing agent and water as the solvent. Starch is a highly abundant biopolymer found in nature, especially in food products such as rice, potato and nuts. This biopolymer can act as a reducing agent in the presence of sodium hydroxide (NaOH), which dissociates in water and the hydroxyl groups (OH^-) react with starch to release electrons that reduce Ag+ to metallic silver. The capping properties of starch are given by its hydroxyl groups, which stabilize the AgNPs through inter- and intra-molecular hydrogen bonding.

In the present work, we report a simple, environmentally friendly, cost-effective and highly reproducible green synthetic approach for the production of colloidal AuNPs, AgNPs and bimetallic Ag/AuNPs with different atomic composition using potato starch as the capping and reducing agent and water as the dispersion medium. The NPs were characterized for their optical properties, colloid stability, particle size, morphology and composition using ultraviolet (UV)–visible spectroscopy, transmission electron microscopy (TEM), X-ray diffraction (XRD), dynamic light scattering (DLS), ζ-potential, energy-dispersive X-ray spectroscopy (EDS) and inductively coupled plasma mass spectrometry (ICP-MS). Antimicrobial studies with commercially available antibiotic-resistant bacterial strains (MRSA and multidrug-resistant [MDR] Escherichia coli) as well as cytocompatibility and anticancer studies with human dermal fibroblast (HDF) and human melanoma cells, respectively, were completed in order to elucidate the biomedical applications of the novel green-synthesized noble metal NPs.

561618 First-Principles Density Functional Theory Modeling and Machine Learning Approach for Redox Potential of Carbon Materials

In this talk, we present the first-principles density functional theory modeling of redox properties of various carbon materials, including quinone derivatives, dopamine, and DNA. First, we discuss the contribution of electron affinity of molecules and solvation to redox potential from our density functional theory (DFT) calculations. Second, we discuss the relationship of redox potential with the design of molecular structure. It is found that aromaticity and electron-withdrawing functional groups can be used to tune the redox potential. Especially, the variation of redox potential as a function of Li association is discussed. The conclusion is that we can tune up the redox potential by designing the detailed molecular structure. Next, all these results are used to build up machine learning. We discuss our artificial neural network and other machine learning algorithms regarding its validation and predictability. Through this study, the redox characteristics of carbon materials are systematically investigated using the first-principles computational modeling method, and then utilized via machine learning methods for high-throughput screening process.

578911 3- D Nanostructures for Next Generation Supercapacitor Application

One-dimensional array architecture such as nanotubes and nanowires has been widely used to enhance the energy storage devices. Supercapacitors have raised the attention over the past decade due to their high-power density, energy density and long life cycle compare to secondary batteries and conventional electrical double-layer capacitors. In this study we are fabricating 3-D structure (Hierarchical) MnO_2 nanowire with high surface area, low toxicity, high abundance, low cost, and environmental friendliness to increase the power density of supercapacitors. Al with less than 1 atom % Cu will be used to form a 3-D porous template during the anodization process. Presence of Cu impurities in an Al film introduces horizontal pores interconnecting the vertically aligned pore structure of the anodized aluminum oxide (AAO) template.

578858 Se-Nanorods Decorated N-Rgonanocomposite for Application in Supercapacitors

It is experimentally proved that selenium and reduced graphene oxide have low specific capacitance due to their chemical nature. Nevertheless, selenium nanorods decorated rGO results in significant increase in the specific capacitance due to formation of the electrochemical double layer at the interfacial area. Moreover, the specific capacitance could be further enhanced by nitrogen doping the rGO sheets. Therefore, novel N-rGO/Se nanocomposite has been successfully synthesized by hydrothermal route starting with graphene oxide, selenium chloride and urea as raw materials. The composite was characterized by XRD, HR-TEM, FESEM, Raman and XPS techniques. The utilized physiochemical characterizations indicated that the final prepared composite can be demonstrated as N-doped rGO decorated by Se nanorods. The supercapacitive performance of the N-rGO/Se nanocomposite has been investigated by cyclic voltammetry and galvanostaticcharge-discharge in 1 M KOH solution and the results reveal that the N-rGO/Se nanocomposite has a large specific capacitance according to CV and galvanostatic charge-discharge tests, while the Se, rGO, and rGO/Se based electrodes has a poor electrochemical performance. Moreover, the as-prepared composite exhibited excellent long cycle stability after 5000 cycles.

579319 HTL Derived Biochar as Electrode Material for Supercapacitors

Hydrothermal liquefaction (HTL) of a lignocellulosic biomass (LCB) produces biochar and aqueous biocrude. After acetone extraction, heavy bio-oil (HBO) is recovered. The leftover biochar is an excellent cost-effective carbonaceous byproduct, characterized by the surface functional groups such as C-O, C=O, and -OH that are dependent on biomass precursors and annealing conditions. Optimization of the annealing conditions could yield high specific surface area (SSA) carbon. In this study, biochar was produced from HTL liquefaction of a LCB at 300°C, followed by annealing at higher temperatures in an inert environment. Following the annealing treatment, the char material was characterized for pore volume, pore diameter, and SSA by BET analyzer and investigated as an electrode for an electrochemical double-layer capacitor (EDLCs) along with Mn:Ti perovskite material. The specific capacitance of the fabricated supercapacitors was measured using cyclic voltammetry. Results obtained with the effect of annealing conditions on the biochar characteristics and charge storage ability of supercapacitors will be presented.

572322 Charge Storage Mechanism of Iron Oxide Electrode Material for Aqueous Electrochemical Energy Storage

The demand for high performance rechargeable batteries is increasing in recent years with the rapid development in electronic devices.

Over the past centuries, iron has been playing an imperative role in shaping human civilization and economic growth, where United States alone generates over 40 million of tons of iron ore per year. There are several advantages making iron an obviously attractive raw material for a rechargeable battery, including low cost, ease of oxidation and multiple oxidation states, a natural and abundant element on earth, and little effects on our environment and human health in its oxide form. Unfortunately, despite the great progress that has been achieved, the performance of iron oxide electrode materials is still unsatisfactory and further improvement is needed.

Our research focus on lepidocrocite (γ-FeOOH) layered materials, which was synthesized via facile one-step wet chemistry. The crystal structure of γ-FeOOH materials was confirmed by X-ray diffraction (XRD) measurement and their electrochemical performance toward aqueous electrochemical storage were tested by Cyclic Voltammetry (CV) in three-electrode half-cell system. The crystal structure change during CV tests was investigated by in situ XRD measurement. Our results showed that the electrochemical performance of γ-FeOOH was dependent on the concentration of sodium sulfate and pH of the electrolyte. Our research showed clear evidence of the energy storage mechanism of iron oxide in alkaline solution, which provides insight on the exploration of new type of charge storage mechanism using iron-based electrode materials.

578484 Potassium Sodium Niobate Based Combined Mode Piezo-Tribo Hybrid Nanogenerator

Piezoelectric and Triboelectric nanogenerators have attained the focus of scientific community due to its ability to convert mechanical energy into usable form of electrical energy which can be used for multipurpose application. In this study, a combined mode Piezo-Tribo Hybrid Nanogenerator has been reported which includes Potassium Sodium Niobate (KNN) based composite including Polyvinylidene Difluoride (PVDF) and Multi Wall Carbon Nanotube (MWCNT) for the Piezoelectric part and Aluminum and Polytetrafluoroethylene (PTFE) induced Polydimethylsiloxane (PDMS) polymer composite as the Triboelectric part. The materials were characterized using Fourier Transform Infrared Spectroscopy (FTIR), Scanning Electron Microscopy (SEM) and X-Ray Diffraction (XRD) techniques. The cylindrical wire like surface of the triboelectric part enables this nanogenerator to operate as a combined mode triboelectric nanogenerator which provided higher triboelectric effect compared to a Vertical mode triboelectric nanogenerator constructed with the same materials. Besides, the piezoelectric layers were tested for different force without poling operation in which higher output was observed for higher force. A maximum 29V and 60V was observed for the piezoelectric and triboelectric layer respectively. This prospective application of this hybrid PTENG for the application in energy harvesting, sensory application and electronic devices has also been discussed.

577154 Data-Driven Tuning of Silicon Alloy Nanomaterials Properties: Novel Insights from Theory and Experiments

Silicon-Germanium alloy materials (SiGe) are one of the principal semiconductors used for high-frequency wireless communication chips, high-speed information communication processors, and modems. With the advent of not only the "Internet of Things" but also 5th generation cellular mobile communications, demand of SiGe semiconductor materials is expected to increase progressively. Silicon alloy materials are also attractive for batteries and energy storage applications. These alloy materials can be synthesized by vapor deposition processes such as Chemical Vapor Deposition (CVD) or Atomic Layer Deposition (ALD) methods to make continuous films, as well as by homogeneous chemical nucleation through pyrolysis to make nanoparticles which can then be tempered to make films. Our study here will focus upon the latter materials synthesis route. However, understanding semiconducting nanomaterials formation from the pyrolysis of mixtures of silane (SiH_4) and germane (GeH_4) under the mildest conditions is still incomplete. It is because homogeneous gas-phase nanomaterials formation is a complex phenomenon in which hundreds to thousands of species undergo simultaneous reactions. The methodology used in our previous work for thermodynamic properties[1] and the kinetics[2] of Si nanoparticles was robust enough to validate the experimental results of pyrolysis of monosilane by Wyller et al.[3] Wyller et al. was able to detect higher order silanes up to pentasilane isomers from the monosilane pyrolysis by developing an advanced gas chromatography/mass spectrometry (GC/MS) method. Further computational modeling can play a significant role in narrowing the gap between controlled experimental studies and practical operating conditions. Automated network generation techniques is one way to describe the kinetics of inorganic nanoparticle formation and investigate all the possible reactions under certain conditions. Rate coefficients are to be estimated for every elementary step comprising the mechanistic model, and kinetic correlations are used to make this tractable. One common method for predicting activation barriers (E_a) is the Evans-Polanyi correlation; however, these structure-activity correlations require detailed thermochemical information for each reacting species. Unfortunately, there are limited studies available that predict the thermochemical properties of SiGe clusters. For this purpose, we conducted a computational study of hydrogenated silicon-germanium alloy clusters ($Si_xGe_yH_z$, $1<X+Y\leq 6$) to predict the thermodynamic and electronic properties.[4] The optimized geometries of the $Si_xGe_yH_z$ clusters were investigated systematically using quantum chemical calculations and statistical thermodynamics. All electronic energies for the clusters were calculated using Gaussian-n methods, in order to validate our approach, we compared our methodology to other popular composite methods such as the CBS-QB3 method and available observed data. Our studies have established trends in thermodynamic properties (standard enthalpy of formation ($\Delta H^o{}_f$), standard entropy (S^o), and heat capacity (C_p)), as a function of cluster composition and structure. Furthermore, we compared HOMO−LUMO energy gaps, and HOMO and LUMO electron distributions in order to gain insight into the electronic stability of the hydrogenated Si, Ge, and SiGe clusters. These properties will be discussed in the context of tailored nanomaterials design and generalized using a machine learning approach.

1. Adamczyk, A. J.; Broadbelt, L. J., Thermochemical Property Estimation of Hydrogenated Silicon Clusters. *J. Phys. Chem. A* **2011,** *115*(32), 8969-8982.
2. Adamczyk, A. J.; Reyniers, M. F.; Marin, G. B.; Broadbelt, L. J., Kinetics of Substituted Silylene Addition and Elimination in Silicon Nanocluster Growth Captured by Group Additivity. *ChemPhysChem* **2010,** *11*(9), 1978-1994.
3. Wyller G. M., Preston T. J., Skare M. O., Klette H., Anjitha S. G, Marstein E. S., Exploring thermal pyrolysis of monosilane through gas chromatography-mass spectrometry measurements of higher order silanes. Silicon for the Chemical and Solar Industry XIV, 2018
4. Choi, Y.; Adamczyk, A.J., Tuning Hydrogenated Silicon, Germanium, and SiGe Nanocluster Properties Using Theoretical Calculations and a Machine Learning Approach. J. Phys. Chem. A 2018, 122, 9851−9868

566518 Understanding the Mechanism of Solvent-Driven Infiltration of Polymer (SIP) into Nanoparticle Packings

We have previously introduced a one-step, room temperature method for the fabrication of high filler-fraction (> 50vol%) polymer nanocomposites through solvent-driven infiltration of polymer (SIP) into nanoparticle (NP) packings. A bilayer thin film of polymer and nanoparticles is exposed to solvent vapor, which leads to capillary condensation of solvent in the voids of the NP packing, and subsequent plasticization and infiltration of the underlying polymer. Herein, we probe how infiltration is driven by a competition between adsorption, solvation and confinement in this system. We have developed an entropic barrier model from self-consistent field theory (SCFT) simulations of a confined, solvated polymer translocating through a narrow pore to unpack the scaling behavior of such systems. We approach this experimentally by tuning the polymer/NP interactions through the use of polystyrene and poly(2-vinylpyridine), which have different interactions with the silica NP surface, and by varying the quality of solvent and molecular weight of polymer. Our understanding of the infiltration dynamics is further informed using coarse-grained molecular dynamics (MD) simulations.

575059 T_g and Structural Recovery of Polystyrene Nanorods

The behavior of glass-forming materials confined at the nanoscale has been of considerable interest over the past two decades with conflicting results sparking debate. Here the focus is on the glass transition and associated structural relaxation kinetics of polystyrene nanorods using the Mettler Toledo Flash differential scanning calorimeter (DSC). The advantages of the Flash DSC include sufficient sensitivity to measure enthalpy recovery for nanogram samples, as well as extension of the measurements to aging times as short as 0.01 s and to aging temperatures as high as 15 K above nominal T_g since high fictive-temperature glass can be created by the fast cooling rates (1000 K/s). The T_g of polystyrene supported nanorods in 20 nm anodized aluminum oxide (AAO) is slightly elevated (~2 K) when compared to the bulk. On the other hand, the T_g of stacked nanorods without support is depressed by as large as 20 K for 20 nm rods at the slowest cooling rates of 0.1 K/s; whereas, T_g is only depressed 10 K at the highest cooling rates of 1000 K/s. These results differ somewhat from those of thin polystyrene films where no T_g depression was observed at the highest cooling rates. Structural recovery is performed as a function of aging time and temperature, and the evolution of the enthalpy is followed. The results will also be discussed in the context of current controversies in the field.

574037 The Effect of Crystallization and Glass Transition Temperature in Thin Poly(D,L-lactic acid) Copolymer Films for Controlling Osteoblast Recruitment and Adhesion

Polylactic acid thin films have significant potential as bioresorbable coatings. The film thickness is also known to affect the transition temperature and crystalline morphology, which is expected to impact cellular adhesion to the coating. Herein, poly(D,L-lactic-co-glycolic acid), PLDG, was spin-coated to yield amorphous films with thicknesses ranging from 30 to 200nm. The amorphous PLDG thin films were annealed at 100°C for 24 hours, 48 hours, and five days and were compared to similar non-annealed samples. Atomic force microscopy, AFM, was used to analyze the morphology of the thin films for indications of crystallization. AFM confirmed that crystallization was apparent on the surface of the film. The crystalline content increased as the annealing time increased from 24 hours to five days. The thickness of the thin films was characterized using ellipsometry. Heat scans from 30°C to 150°C were performed on the ellipsometer to determine the linear expansion coefficient as a function of temperature. The linear expansion coefficient is lower below the glass transition temperature and significantly higher above the glass transition temperature. For example, for a 212 nm film, the linear expansion coefficient below the glass transition is 4.23×10^{-4} °C^{-1} and above the glass transition the linear expansion coefficient is 2.03×10^{-3} °C^{-1}. In these scans, both the glass transition and melt temperatures could be clearly identified. Results reveal that the glass transition and melt temperature in thin PLDG films were lower than in bulk samples of PLDG. Control of the crystallization may help promote better adhesion of osteoblast cells to thin films. Future work will investigate the effect of crystallization on the degradation of thin films and osteoinductivity of PLDG thin films.

572580 Dynamic Phase Transitions in Confined Polymer Glasses

Despite more than two decades of effort, several qualitative features regarding the segmental dynamics of thin films of glass-forming materials remainly poorly understood. This challenge is of paramount importance in thin film membranes, organic electronic devices, and critical steps of semiconductor manufacturing. In this talk, I will present the results of recent simulations and experiments that suggest the dynamics near surfaces are qualitatively different from those of a bulk supercooled liquid, suggesting that the dynamics near a free surface lose their glassy character. To explore this hypothesis farther, we have used molecular simulations to investigate a non-equilibrium phase transition associated with glass-forming materials. This transition is associated with a supercooled liquid that undergoes a sharp transition from a mobile dynamic phase to a phase with reduced mobility, and it provides a clear signal in the dynamics of the glass-forming polymer under confinement. I will describe our characterization of this transition in free-standing thin films of supercooled liquids, where we find that only the bulk of the film is able to undergo this transition. This suggests that the dynamics at the free surface are not glassy in nature, at least not at the temperatures accessible to our simulations.

570629 Effect of Molecular Structure and Crosslink Density of Surface-Anchored Poly(N-alkylacrylamide) Hydrogels on Adsorption of Fibrinogen

We describe a simple approach to generate surface- attached biocompatible hydrogels with tunable cross-link density and employ them to study the effect of gel structure on protein adsorption. Using free-radical polymerization, we synthesize a series of random copolymers comprising N-alkylacrylamide and the photoactive curing agent Methacryloyl-oxy-benzophenone (MABP) of mole fractions ranging from 2.5 to 10%. We deposit a thin film of the precursor copolymer (~150 nm) on a silicon or glass substrate, which is precoated with monolayers of benzophenone–silane, then cross-link it through UV irradiation at 365 nm (dose ≈ 6–10 J/cm^2) to generate surface-attached networks. A systematic investigation of the network properties such as gel fraction, cross-link density, and swelling ratio reveals that gels with higher MABP content (≥5%) and longer alkyl chain produce densely cross-linked hydrophobic networks with low or no swelling in an aqueous medium. We study the adsorption of fibrinogen (Fg) on such hydrogel substrates and establish that the amount of adsorbed Fg depends on the degree of cross-linking, hydrophobicity and the swelling capacity of the networks. Specifically, although Fg adsorbs heavily on denser networks, loosely bound gels that swell in aqueous medium repel proteins. We attribute the latter behavior to entropic shielding and size-exclusion factors.

567278 Probing the Interface of Irreversible Adsorption in Polymer Nanocomposites

Annealing above the glass transition temperature (T_g) can prompt the formation of irreversibly adsorbed layers at polymer-substrate interfaces, which can subsequently influence the local and bulk properties of polymeric systems. This irreversible adsorption is predicted to be particularly influential in polymer nanocomposites due to their high processing temperatures and substantial interfacial area. However, until now the local and bulk effects of irreversible adsorption in polymer nanocomposites have not been directly described. Herein, we consider the effects of irreversible adsorption in polystyrene-silica nanocomposites using conventional bulk DSC measurements as well as direct, local techniques such as fluorescence spectroscopy to illuminate the relationship between local T_g perturbances proceeding from irreversible adsorption and overall nanocomposite properties. We also characterize the growth of the adsorbed layers as a function of annealing time. This characterization of local interfacial dynamics provides insights necessary for the engineering of polymer nanocomposite materials.

578251 Lattice Self-Consistent Field Calculations of Ring Polymer Brushes

We report the first systematic study using lattice self-consistent field (LSCF) calculations of ring homopolymer brushes grafted onto a flat and homogeneous surface and immersed in an explicit and athermal solvent, which are either uncompressed, compressed by a flat and impenetrable surface, or compressed by an identical brush.[1] Our results clearly show that ring brushes are slightly less stretched than, thus nearly but not completely identical to, the "equivalent" linear brushes having half the chain length and double the grafting density. Our LSCF results are consistent with the molecular simulation results reported in the literature[2,3], except the previous finding that the normal pressure of two opposing ring brushes is only half of the "equivalent" linear brushes at melt density[3].

[1] Qiu et al., *Soft Matter* **2018**, *14*, 1887.

[2] Reith et al., *Europhys. Lett.* **2011,** *95*, 28003.

[3] Erbas and Paturej, *Soft Matter* **2015,** *11*, 3139.

576724 Align Graphene Layers Using Polymer Interfacial Interactions

Inter- and intra-molecular interactions such as coiling or polymer chain entanglements can generate stresses or strains of different degrees. These forces or displacement as a result of polymeric diffusion or stretching can be used for nanomaterial synthesis. For example, the mass production of high-quality graphene has been challenging, especially when the in-plane morphologies of these exfoliated sheets need to be controlled in optics, electronics or composites. In addition, as compared to their allotropes of carbon nanotubes or fullerene, graphene exhibits higher mechanical robustness of modulus of ~1 TPa and strength of ~130 GPa. However, these intrinsic properties have not been achieved in polymeric composites due to the complexity of graphene conformations within the molecules. This research aimed at achieving aligned and exfoliated graphene from stacked graphite layers using interfacial interactions among different phases in composites. The polymer-graphene interactions going beyond the graphene-graphene Van der Waals forces will lead to the aligned graphene layers laminated between polymer channels. Our design of multi-phase spinneret allowed the fabrication of coaxial polymer/graphene/polymer composite fibers. A stepwise drawing of the multiphase samples produced fibers with different dimensions and molecular morphologies; more importantly, the graphene will be exfoliated and aligned due to the shear stress by the comparative movements of different polymer channels. Significant mechanical and electrical properties were induced due to the graphene layer reductions and orientations. X-ray diffraction (XRD), scanning electron microscopy (SEM), differential scanning calorimetry (DSC), Raman, and, thermogravimetric analysis (TGA) were used to characterize the material structure and properties.

571886 Shape Memory Assisted Contact Printing in Nanoscience and Electronics

Over the past two decades, contact transfer printing has evolved into a viable manufacturing technology that can deposit and pattern various organic, polymeric and inorganic ink materials with micro- and nano-scale precision. Contact printing relies on the elastomeric stamp that forms an adhesive and conformal contact with the ink layer during the ink pickup, and on the interfacial fracture of the ink-stamp interface during the ink transfer. Contact printing is inherently amenable to replicate large-area patterns on flat or curvilinear substrates, and it has potential to evolve into a universal platform for large-area, parallel deposition of multiple types of materials at the sub-micrometer length scale. However, the key to enabling such manufacturing is to establish clean and reliable methods for controlling interfacial adhesion and fracture mechanics during the ink pickup and release.

In the past, we have demonstrated that the adhesive stamp-ink interactions can be controlled by the chemical composition and stiffness of the polymeric stamps. For example, the surface energy of the polymers can be controlled chemically, producing stamps with tunable polarity. As a consequence, high and low surface energy stamps can be used to pattern a variety of electronic materials with ~100nm resolution and high uniformity. More recently, we have shown that modulation of the interfacial adhesion can be achieved with stamps made of shape-memory polymers that can change their contact area with inks using external stimuli. We have demonstrated that by modulating the stamp adhesion using thermo-mechanical cycles, we can pick up and deposit multicomponent films of inorganic and organic electronic materials from the donor substrate to the receiver surface. Our approach permits patterning of multilayered stacks of organic/inorganic thin films with <1μm resolution and high uniformity. We envison, that the developed method will find numerous applications in nanoscience and electronics. For example, we are currently adapting our technology to pattern complete pixels of the μ-LED devices with sub-micrometer resolution using a unified set of materials and printing conditions.

574980 Electrochemical Growth of Charge Transfer Complex Micro/Nanowires on Microelectrodes and Their Application in Gas Sensing

Nanowire sensors are promising to deliver greater performance than conventional sensors due to their size confinement as well as low cost and power consumption. However, scalable assembly of nanowire sensors remains a big challenge, which requires the synthesis of uniform nanowires and organization of them. Here, we report a simple method to control the growth and organization of a promising nanowire material for gas sensing, tetrathiafulvalene bromide ((TTF)Br), by mass transfer in electrodeposition. We found that the use of microdisk electrodes for electrodeposition of (TTF)Br wires resulted in wires with higher uniformity and larger aspect ratio (length to width) than their counterparts deposited on films under the same deposition conditions. More interestingly, we found that the number of wires grown from a microdisk electrode exhibited a limiting number of 3 or 4 per electrode when the electrode diameters are smaller than ~1.4 μm. Both experimental and simulation results suggest that the morphological differences and the constant minimum wire density are caused by the different TTF flux distributions at a microdisk versus a film electrode. These findings will be used to show a scalable manufacturing method of nanowire-based sensing devices for gas detection.

556998 Process Safety and Scale-up of Graphene Oxide Synthesis and Storage

Graphene oxide (GO) has been demonstrated as an effective additive for energy storage and composite applications, and large-scale production of GO is of increasing commercial interest. However, a number of process safety issues need to be addressed before scaling up GO production. In this work, the safety issues with GO production are addressed in two studies. The first study addresses the exothermicity of the oxidation reaction during modified Hummer's Method. The oxidation reaction is studied in a Reaction Calorimeter, and the data show that the heat of reaction of the system is capable of reaching the decomposition temperature of manganese heptoxide. The second study focuses on the thermal stability of as-produced GO. The explosive decomposition of GO is studied in an Advanced Reactive System Screening Tool (ARSST). The data indicate that the explosive decomposition temperature of GO strongly depends on sample size, and pressure generation can reach 1000s of psi per minute for less than a gram of material. These results are beneficial in evaluating safer operating conditions for synthesis, post-processing, and storage of GO, particularly in large batches.

573243 Laser Ablation Synthesis in Solution for the Rational Design of Hybrid Carbon-Based Nanocomposites for Enhancements in Electrochemical Storage and Conversion Systems

The increasing global demand for clean and sustainable energy sources has given rise to the significant interest in the rational synthesis and design of low-cost and efficient functional nanocomposites and nanomaterials as catalyst and/or supercapacitors in advanced electrochemical energy storage/conversion systems. Nonetheless, the design of such nanomaterials with tailored interfacial functionality and charge transport properties relies on systematic and fundamental understanding of the processing-structure-property relations for precise tailoring of architectures and compositions of the active nanomaterials without using any undesired chemicals/surfactants/ligands that can compromise their electrochemical performance and interfacial activities . Herein, we present our most recent findings in using Laser Ablation Synthesis in Solution (LASiS) as a strategy and pathway for the facile and environment-friendly, yet efficient, synthesis of new classes of carbon-based hybrid nanocomposites (HNCs) decorated with metal-oxide nanoparticles (NPs) that exhibit unique electrocatalytic and supercapacitive properties. Initially, we report the use of LASiS coupled with two-post treatments for the fabrication of (1) Co_3O_4 NPs/reduced graphene oxide (rGO), (2) Co_3O_4 nanorods (NR)/rGO, and (3) Co_3O_4 NPs/nitrogen-doped graphene oxide (NGO). This methodology allows us to tune the selective functionalities of the HNC by adjusting their structure-property relationships. The Nitrogen doping in the NP/NGO HNC, for instance, promotes a higher electron conductivity while enhancing the electrochemical activity and preventing the aggregation between NPs. These interfacial energetics and arrangements have indicated to yield superior Oxygen Reduction Reaction (ORR) electrocatalytic activities. Similarly, the NR/rGO HNCs present a network of interconnecting 1D nanostructures that give rise to an effective charge transport and electrolyte diffusion at the electrode-electrolyte interfaces, resulting in the enhancement of the supercapacitive properties. On the other hand, the NP/rGO HNCs demonstrated intermediate functionalities towards both ORR catalysis and supercapacitance.

558589 Micro-Chemical Technology Tune Nanoscale Structure of Nanoparticles and Its Bioapplication

Nanoparticles have been extensively studied due to their wide application, especially in the area of bio-medicine, including drug-carrier, targeting delivery, imaging, etc. Recently, the systematic regulation of nanostructure has attracted considerable research interests, as it is found that to achieve optimum accumulation of nanotherapeutics specifically at diseased sites, not only the sizes but the nanoscale structure also need to be optimized. In this study, the micro-structure of nanomaterials (including external structure, internal structure) have been systematically tuned by a micro-chemical technology, named Flash Nanoprecipitation (FNP). Through tuning the polymer concentration, the external structure (shape) could be controlled from nano-rod, to micro-rod and nano-sphere. *In vivo* experiment exhibit that the nanorod particles with a polymer shell display excellent tumor targeting capability. In addition, through tuning stream velocity, the internal structure could also be controlled from compact to loose formation. *In vitro* drug release show that the compact formation has a relatively lower release velocity. Therapeutic and imaging applications may therefore benefit from the strategy adopted here.

High-Density Oxygen Storage in Nanometer-Scale Deep Capillaries

Adam Duzik, Greg Chester, Justin Hill†

Mainstream Engineering Corporation, Rockledge, FL, USA

†Corresponding author: jhill@mainstream-engr.com

Abstract: We present a general gas storage system that uses nanocapillary arrays to replace conventional high-pressure gas cylinders. Aluminum anodization produces a regular array of nanocapillaries with ultra-high aspect ratios that can be filled with aligned, highly stable carbon nanotubes (CNTs) by chemical vapor deposition (CVD). By sealing the nanocapillaries with a membrane electrode-sealing assembly (MESA), gas can be electrochemically pumped into and out of these nanocapillaries. By stacking numerous of these nanocapillary arrays as a storage device, a partial rupture vents only the gas in the individual array, preventing total containment failure, and creating a safe approach to achieve high-pressure gas storage. The nanocapillary approach is also theoretically capable of 2–3 times the gravimetric and volumetric storage capacity of compressed gas cylinders.

Many parameters contribute to storage capacity, primarily: nanocapillary depth and MESA pressure tolerance. Current progress shows capillary depths up to approximately 3.1 mm at an average pore diameter of 100 nm. Cyclical polarization results indicate electrochemical pressurization via reversible dissociation of input species into ionic form for transport through the ionomer in the MESA. Various combinations of internal and surface catalysts were explored for pumping specific gas chemistries. Measurements showed storage and release capability for up to 10 cycles.

This paper cover the required advances in nanocapillary storage device manufacturing as well as the characteristics of electrochemical compression in nanocapillary arrays. Overall storage device efficacy will be discussed.

Introduction

Current self-contained breathing apparatus (SCBA) devices support three primary sources for oxygen supply: compressed or liquid oxygen, chemical oxygen systems, and water-based oxygen-generating systems. These approach tends to be unsafe, bulky or heavy, and are limited in supply rate, require complex logic devices, or are generally awkward for SCBA applications. Mainstream has developed an alternative oxygen storage system based on anodized aluminum oxide (AAO) nanocapillaries which are sealed by a gas diffusion membrane allowing them to be pressurized with oxygen from a humidified air stream. In this system, oxygen is electrochemically compressed into the AAO nanocapillaries, which have an extremely high pressure rupture limit owing to the nanoscale pore radius and the symbiotic strength of neighboring pressurized pores. The ratio of the nanocapillary wall thickness to diameter (and thus pressure tolerance) is highly tunable. Furthermore, the pore wall is coated with a high-tensile-strength material, enabling maximum gravimetric (G_c) and volumetric (V_c) storage densities. The nanocapillary density is approximately $2.5*10^9$ capillaries/cm^2. Thus the nanocapillary array has the potential to store O_2 at higher pressures than the current compressed gas cylinders. In addition, total containment failure due to side wall rupture of a gas storage cylinder is mitigated because only a small fraction of the total gas storage is stored within each individual nanocapillary.

Compressing O_2 our approach only requires electrical power and either water or humidified ambient air at the filling station. Furthermore, the system, including gas regulation and delivery within the SCBA unit itself, requires no moving parts, decreasing the maintenance and dramatically reducing the logistical burden of transporting high-pressure O_2 or unstable chemically stored O_2 sources to forward operating areas.

Our nanocapillary-based oxygen storage system consists of four main components:

1. An AAO structure consisting of nanocapillaries electrochemically formed on aluminium sheets.[1]
2. CNTs in the capillaries that act as wall reinforcement as well as forming the inner electrode for the eelctrochemcial compression system.[2,3]
3. An ion exchange membrane to selectively transfer ions into and out of the nanocapillaries. Futhermore, the membrane acts to seal stored gas in the nanocapillaries.
4. A catalyst inside the capillaries and on the membrane surface to facilitate oxygen reduction and evolution during pump in and pump out.

Items 3 and 4 are the major focus of this work. In this configuration, we refer to the membrane/cap as a MESA because of its ability to both act as the seal and electrode for the compression of O_2 within the nanocapillaries. Catalyst selection builds on literature investigations into O_2 evolving catalysts, primarily IrO_x.[4–7] The nanocapillary array fabrication process starts with high-purity Al which is anodized to produce deep pores as shown in Figure 1a. CNTs were deposited by CVD, into the capillaries to provide the inner electrode in the MESA and base as shown in Figure 1b.[2,3,8]

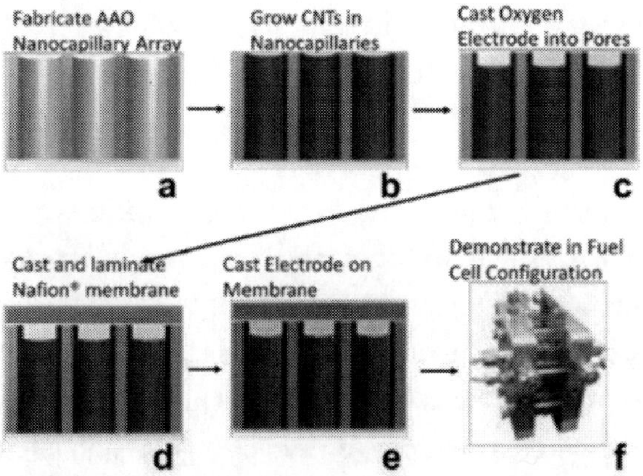

Figure 1. Outline of fabrication procedure for nanocapillary arrays,depositing the CNTs, mebrane and catalyst layers to fabricate membrane electrode seal assembly.

After this, the inner CNT electrode is applied to make electrical contact with the Aluminium base and line the pore This is followed by spraying a mixture of oxygen-reducing and oxygen-evolving catalysts onto the inner electrode, as shown in Figure 1c.

Two final important components remain: a membrane seal and the outer catalyst layer. This work focused on selection and optimization of the membrane and catalyst. The former maintains the O_2 pressure within the capillaries while the latter facilitates the electrochemical reactions required for the dissociation or reduction of oxygen and the subsequent evolution of oxygen. For the ion exchange membrane, we chose Nafion®, which is a high-conductivity ionomer for transferring H^+. Nafion®. It was first sprayed over the inner electrodes (Figure 1d) to ensure good ionic contact, and then the nanocapillary array was capped with a Nafion® membrane film and an electrode impregnated with a Pt catalyst. This whole assembly was compressed in a hot press to laminate the structure and make the completed device ready for testing (Figure 1e–f).

The pump-in operation of the finished MESA device is shown in Figure 2. Voltage polarity between the electrodes is used to control the pumping direction. The pump in, or oxygen storage part of the cycle, is described first. The applied voltage is negative at the outer electrode and positive on the inner capillary electrodes (defined as forward bias). Water vapor molecules stored in the nanocapillaries are electrochemically split into H^+ ions (protons) and O_2 molecules. This is defined as the oxygen evolution reaction (OER). The resulting oxygen is stored within the capillary and the H^+ ions are transported through the Nafion® membrane. At the outer electrode, oxygen from the ambient air feed is reduced and combines with the protons to form water. This is the oxygen reduction reaction (ORR). These reactions combine to increases the O_2 pressure inside the capillaries. Since Nafion® is also permeable to H_2O, both by diffusion and electroosmotic drag with the electromigrating protons, humidified input air must be used to act as the water source, and to suppress H_2O loss from the cell. This is vital as the oxygen from the H_2O molecules in the nanocapillaries act as the oxygen source for the stored O_2; losing water to diffusion will render the MESA ineffective.

The polarity of the electrodes can be reversed to controllably release the oxygen (inner electrode negative, outer electrode positive). This allows the electrolysis of the H_2O molecules in the outer gas stream into H^+ ions and O_2 molecules. The H^+ ions diffuse through the Nafion® membrane, recombining with the stored O_2 which are reduced at the inner electrode and converted into H_2O molecules. This is the oxygen evolution reaction (OER).

Only a few volts are required to drive the pressurization process, making it low-energy. Furthermore the applied voltage or current can be used to control the desired O_2 flux (fill or delivery rate).

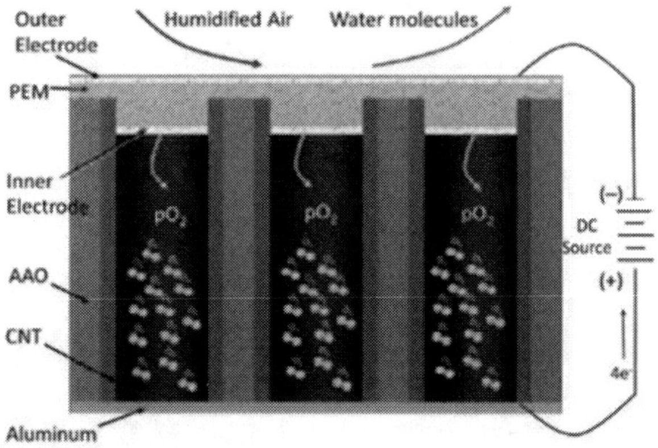

Figure 2. O_2 compression and generating system integrated into the AAO nanocapillary array.

Experimental Procedure

The inner and outer electrodes of the MESA must operate in both an OER and ORR mode and thus must be optimized for both reactions. However, since these electrode reactions require different electrode characteristics we needed to investigate catalyst and electrode material selection which would work well for both reactions. We fabricated several samples using various catalyst mixtures to identify the most effective combination.

We identified catalysts for use with the MESA including Pt for the ORR and IrO_x for the OER. The reason the subscript "*x*" is used is that the exact oxidation state of the iridium is likely a mixture of +2, +3, and +4 charge states. To synthesize/deposit the IrO_x catalyst into the pores without blocking the pores, we explored a method to synthesize 2 nm IrO_x catalyst particles and prevent aggregation of the particles without using a ligand, which could limit catalytic performance.[12,13]We tested the catalytic performance of these IrO_x nanoparticles both alone and in combination with platinum impregnated into the CNT–AAO templates. We prepared three CNT-coated templates and tested the catalytic performance in 0.5 M H_2SO_4 saturated with O_2 gas. The testing was conducted in a three-electrode setup on a glassy carbon support as well as with the CNT-coated template with or without spray cast catalyst acting as the working electrode. We used a platinum-coated titanium mesh as the counter electrode and a saturated calomel electrode as the reference electrode. Testing was conducted at room temperature while stirring at 400 rpm. The catalytic performance was evaluated by Cyclic voltammetry (CV) with a sweep rate of 100 mV/s conducted for 3 cycles. All potentials are reported versus a reversible hydrogen electrode (RHE) and are iR-corrected using resistances collected from high-frequency impedance measurements.

We tested three templates with different catalyst mixtures. One template was coated by spray casting with 1.35 mg of Pt mixed with Nafion at a mass ratio of 0.18:1 Nafion solids to Pt. Similarly, 1.35 mg (metal basis) of IrO_x mixed with Nafion was spray cast on a second template. Finally, a third template was cast with Pt and IrO_x at a mass ratio of 0.18:1:1 Nafion to Pt to IrO_x. The resulting CV scans for each sample are plotted in Figure 3 and tabulated in Table 1. The standard potentials of ORR vs. RHE (0.00 V) and OER vs. RHE (1.23 V) are indicated by the vertical dotted lines.

Results

Cyclic voltammetry of catalyst coated CNTs

When testing the catalytic performance of Pt and IrO_x, a. We first showed that uncoated CNTs exhibit little catalytic activity for either OER or ORR. Tests in 0.5 M H_2SO_4 on bare CNTs with and without the catalysts are shown in Figure 3.

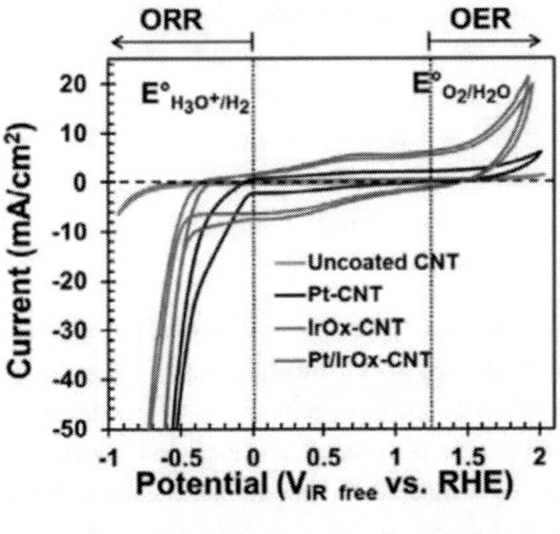

Figure 3. Cyclic voltammetry of the OER region of varying catalysts in 0.5M H_2SO_4 using a sweep rate of 100 mV/s; corresponds to the numbers in Table 1.

Table 1. Catalytic Performance for OER and ORR in 0.5M H_2SO_4.

Catalysts	Overpotential (V_{iR_free}) Relative to:		Potential at Specified Current Density (V_{iR_free} vs. RHE)	
	ORR: $E^{\circ}_{H_3O^+/O_2}$ (0.00 V vs. NHE)	OER: $E^{\circ}_{H_2O/O_2}$ (1.23 V vs. NHE)	ORR: −15 mA/cm²	OER: 15 mA/cm²
CNT	0.83	> 0.75	< −1.00	> 2.00
Pt–CNT	**0.03**	0.55	**−0.27**	> 2.00
IrO_x–CNT	0.50	0.26	−0.55	1.84
Pt/IrO_x–CNT	0.46	**0.24**	−0.49	**1.80**

The Pt-only catalyst performed only marginally better for the OER at high overpotential but showed a much higher ORR catalytic activity and a lower overpotential. In contrast, IrO_x and the Pt–IrO_x mixture displayed much lower overpotential and higher catalytic activity for the OER, although these had higher overpotentials and similar catalytic activity for the ORR as shown in Figure 3 and Table 1. Interestingly, the Pt–IrO_x mixture exhibited a lower overpotential for OER than IrO_x alone, but displayed good performance for ORR and the best performance for OER. Based on these results, we selected Pt–IrO_x as the catalyst of choice; however, we still needed to determine the best Pt–IrO_x composition to spray into the nanocapillaries.

Catalyst Composition Optimization in the Nanoporous CNT Templates

We used cyclic voltammetry to compare the catalytic performance of uncoated CNTs, plain Nafion®, and various catalyst compositions for the both the OER and ORR. The uncoated CNT sample (no Nafion® or catalyst) exhibited almost no catalysis for OER and had very high overpotential for ORR as expected. However, the Nafion-only sample exhibited the same catalytic performance as Pt or 1:1 Pt:IrO_x mixtures. This result is suspicious because the Nafion-only CV curve should be nearly the same as the uncoated CNT, which is not catalytic for either reaction direction. This result could indicate contamination of the spray process sample, as leaving behind even trace amounts of Pt in the spray equipment is sufficient to produce a significant increase in the activity. As can be seen in Figure 4, the 1:2 Pt:IrO_x catalyst ratio showed the lowest overpotential and highest catalytic current response. With the optimized catalyst mixture identified, simulation runs of real-world operation were carried out.

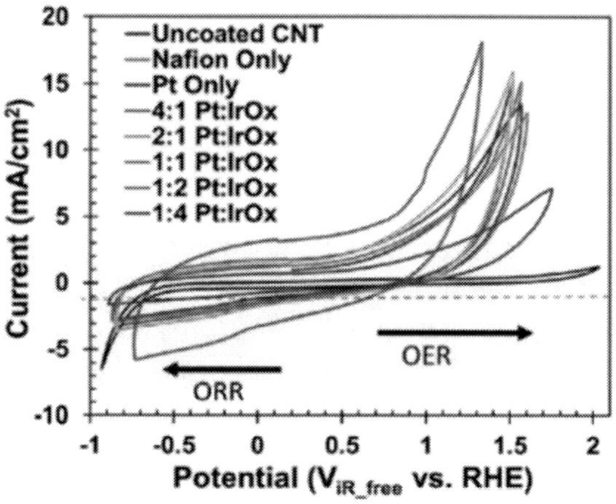

Figure 4. Cyclic voltammetry of varying catalysts on CNT AAO templates in 0.5 M H_2SO_4 using a sweep rate of 100 mV/s.

Full Cell Tests

The SCBA oxygen delivery system will need to withstand many pump-in and pump-out cycles, stressing the catalyst layer in both directions (ORR-OER) over many cycles. As such, we performed repeated CV cycling on the 1:2 Pt:IrOx sample and measured performance changes throughout the process, as shown in Figure 5. We ran the first 12 CV cycles in one day and performed the final nine after resting the mixture overnight to determine the non-cycling stability. The CV maximums show that the sample was equally functional after resting overnight and equilibrating within the system. These equal starting points were likely due to overnight leaking of any stored O2 leading to an empty storage structure when testing resumed. During CV cycling, variation of the O2 concentration, as measured by an O2 sensor after the storage cell, indicated O2 storage and release was occurring.

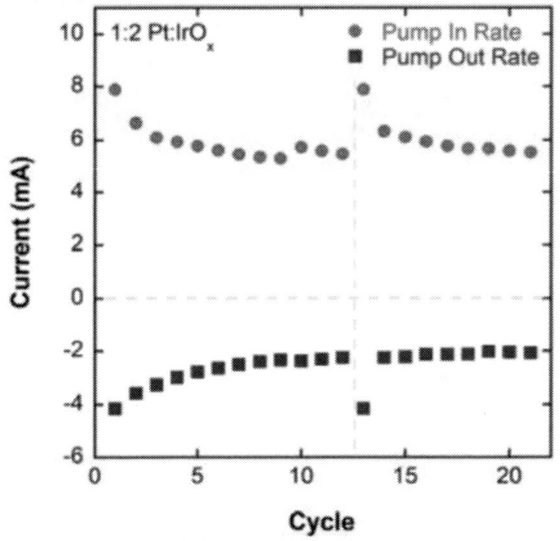

Figure 5. The 1:2 Pt:IrO$_x$ sample with high activity and prolonged activity through at least 21 CV cycles.

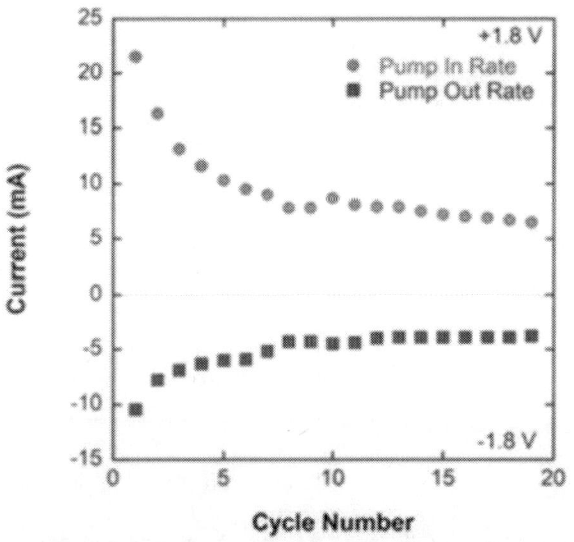

Figure 6. CV peak current for cycling 1:2 Pt:IrO$_x$ catalyst composition. O_2 pressurization showed decreasing activity throughout the 19 cycles. O_2 releasing showed no change after 12 cycles, indicating catalyst breakdown.

We performed a second set of tests to confirm the cyclic stability of the 1:2 Pt:IrO$_x$ catalyst composition. The positive and negative CV peaks correspond to pumping into and out of the prototype, respectively. Both pumping in and out were at their highest during the first cycle and then decreased throughout testing as seen in Figure 6. Pumping out showed no change in performance after the 12th cycle, which likely indicates the catalyst had ceased to function properly either due to unfavorable thermodynamic conditions stemming from the stored O_2 content or catalyst degradation. We made a similar sample without the catalyst spray before Nafion® capping to show the role of the inner catalyst. The CV results for the single catalyst sample are shown in Figure 7. The pump in rate for the first cycle was as before because the outer catalyst was still present, but the pump out rate was much lower, indicating little to no O_2 reduction inside the capillaries.

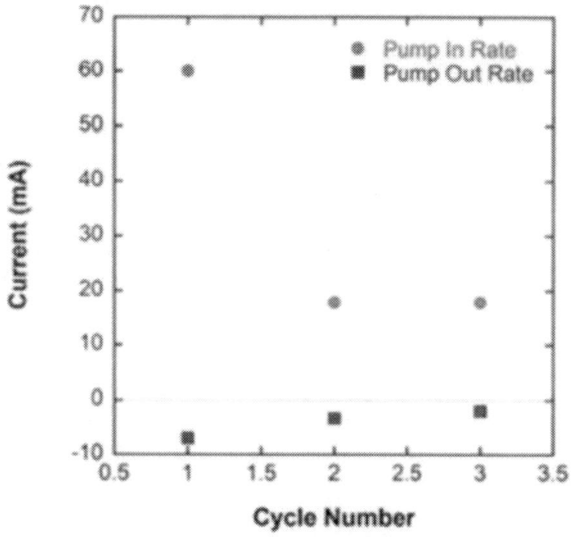

Figure 7. CV peaks for an O_2 storage prototype with no inner catalyst, showing reduced ability to pump in after the initial cycle and no pumping out.

Conclusions

In this work we investigated new O_2 storage techniques using AAO nanocapillaries, CNTs, and a Nafion® membrane as a means of transferring O_2 into and storing O_2 in the nanocapillaries. Pt and IrO_x catalyst mixtures were used to dissociate O_2 with low applied voltage. Cyclic voltammetry sweeps were done on Pt:IrO_x catalyst mixtures with atomic ratio of 4:1, 2:1, 1:1, 1:2, and 1:4. The 1:2 Pt:IrO_x catalyst mixture was shown to be the most effective, producing the highest current responses and thus optimal O_2 evolution. This result demonstrated the viability to store O_2 in a nanocapillary array. Power requirements to electrochemically pump in and extract usable O_2 were low. However, repeated cycling showed catalyst breakdown and degradation after about 11 cycles. More work is needed to find alternative catalysts and catalyst application techniques capable of withstanding many more cycles.

Acknowledgements

This material is based upon work supported by the Army Contract Command under Contract No. W911SR-16-C-0036. Any opinions, findings and conclusions or recommendations expressed in this material are those of the author(s) and do not necessarily reflect the views of the Army Contract Command. DISTRIBUTION A. Approved for public release: distribution unlimited.

References

1 Chahroura, K. M., Ahmed, N. M., Hashim, M. R., Elfadill, N. G., Al-Diabat, A., M., Bououdina, M. Influence of wet etching time cycles on morphology features of thin porous Anodic Aluminum oxide (AAO) template for nanostructure's synthesis *Journal of Physics and Chemistry of Solids*, **87**, 1–8 (2015).

2 Yuan, Z., Huang, H., Liu, L. & Fan, S. Controlled growth of carbon nanotubes in diameter and shape using template-synthesis method, *Chem. Phys. Lett.*, **345**, 39–43 (2001).

3 Hu, W., Gong, D., and Chen, Z. Growth of well-aligned carbon nanotube arrays on silicon substrates using porous alumina film as a nanotemplate, *App. Phys. Lett.*, **79**, 3083–3085 (2001).

4 Chang, C., Yuen, T., Nagao, Y. & Yugami, H. Electrocatalytic activity of iridium oxide nanoparticles coated on carbon for oxygen reduction as cathode catalyst in polymer electrolyte fuel cell, *Journal of Power Sources*, **195**, 5938–5941 (2010).

5 McCrory, C. C., Jung, S., Peters, J. C. & Jaramillo, T. F. Benchmarking Heterogeneous Electrocatalysts for the Oxygen Evolution Reaction, *Journal of the American Chemical Society*, **135**, 16977–16987 (2013).

6 Han, X., Cheng, F., Zhang, T., Yang, J., Hu, Y., & Chen, J. Hydrogenated Uniform Pt Clusters Supported on Porous $CaMnO_3$ as a Bifunctional Electrocatalyst for Enhanced Oxygen Reduction and Evolution *Advanced Materials*, **26**, 2047–2051 (2014).

7 Chang, C., Yuen, T., Nagao, Y. & Yugami, H. Catalytic activity of carbon-supported iridium oxide for oxygen reduction reaction as a Pt-free catalyst in polymer electrolyte fuel cell, *Solid State Ionics*, **197**, 49–51 (2011).

8 Zhou, L. Progress and problems in hydrogen storage methods, *Ren. Sust. En. Rev.*, **9**, 395–408 (2005).

9 Huczko, A. Synthesis of aligned carbon nanotubes, *App. Phys., A***74** (2002).

10 Segura, R. A., Ibáñez, W., Soto, R., Hevia, S. & Häberle, P. Growth morphology and spectroscopy of multiwall carbon nanotubes synthesized by pyrolysis of iron phthalocyanine, *J. Nanosci. Nanotech.*, **6**, 1945 (2006).

11 Xu, Y., Flor, E., Kim, M. J., Hamadani, B., Schmidt, H., Smalley, R. E., Hauge, R. H. Vertical Array Growth of Small Diameter Single-Walled Carbon Nanotubes, *J. Am. Chem. Soc.*, **128**, 6560–6561, doi:10.1021/ja060944+ (2006).

12 Nakagawa, T., Beasley, C. A. & Murray, R. W. Efficient Electro-Oxidation of Water near Its Reversible Potential by a Mesoporous IrOx Nanoparticle Film, *The Journal of Physical Chemistry C*, **113**, 12958–12961 (2009).

13 Zhao, Y., Hernandez-Pagan, E. A., Vargas-Barbosa, N. M., Dysart, J. L. & Mallouk, T. E. A high yield synthesis of ligand-free iridium oxide nanoparticles with high electrocatalytic activity, *The Journal of Physical Chemistry Letters*, **2**, 402–406 (2011).

571201 First-Principles Computational and Machine Learning Approach to Innovative Design of Electrocatalysts

plays a key role in realization of various energy devices and efficient production of industrial goods. This presentation introduces frontiers of research works for the objective using the first principles-based multiscale computational framework combined with machine learning technique. As the example, nanoscale electrocatalysts in proton exchange membrane fuel cells (PEMFCs) are introduced. It clearly suggests that the method provide new design principles beyond conventional wisdom through identification of fundamental descriptor and mechanisms on the catalysis. Furthermore, combined with machine learning technique the method even can screen best electrocatalysts among a wide range of candidates, eventually rendering a computational materials genome approach to substantially reduce temporal and material cost necessary for the cycle from discovery to commercialization into markets of a new electrocatalyst.

568440 Improving Precision in Materials Synthesis for Hydrogen Electrocatalysis

Our research group is broadly interested in the synthesis and characterization of nanoparticulate catalysts for electrochemical energy conversion in aqueous alkaline electrolytes. Our primary focus to date has been on nonprecious catalysts for hydrogen evolution/oxidation in the long-term pursuit of efficient renewable energy storage via alkaline unitized regenerative fuel cells. This field has gained increased interest recently due to the attractiveness of alkaline operating conditions for capital cost reductions and the emerging availability of high-performing oxygen evolution/reduction catalysts and alkaline anion exchange membranes. By contrast, work on alkaline hydrogen electrochemistry has been hindered by two research challenges. The first is the lack of consensus regarding materials and methods for reproducible controls against which new catalysts can be evaluated. The second is the difficulty of making "apples-to-apples" comparisons between novel catalyst candidates that exhibit widely different morphologies. This presentation will describe our recent work to address each of these challenges by deploying precise synthetic methods for nanocatalysts.

To address the challenge of enabling high-quality controls in hydrogen electrochemistry, we have developed a straightforward method for generating clean Pt nanoparticles. The approach is based on a simple thermal reaction between a Pt(IV) precursor and ascorbic acid in the presence of sodium polyacrylate as a solubilizing agent. The resulting polycrystalline nanoparticles are uniform and 3-4 nm in size which can be readily recovered and cleaned using an aqueous base treatment. They can be directly deposited on any substrate of interest with or without a catalyst support, and they demonstrate consistently high activity toward hydrogen evolution/oxidation in acid and alkaline electrolytes.

In the interest of facilitating direct comparisons between candidate hydrogen evolution/oxidation electrocatalysts, we are leveraging a versatile method for synthesizing metal composite nanoparticles that involves the use of extremely powerful reducing agents in nonaqueous solution. This method enables us to generate uniform nanoparticles comprising mixtures of nearly any mixture of transition metals. We are also developing a unique technique adapted from the semiconductor nanofabrication literature to synthesize polycrystalline thin films whose surface roughness is on the order of a single bond length. This degree of dimensional control is expected to enable new insights into electrocatalyst structural evolution under operating conditions.

574668 Nanoscopic Control of the Interfaces to Optimize Metal-Insulator-Semiconductor Systems for Solar Water Splitting

Harnessing sunlight to split water is a promising route to produce sustainable hydrogen fuel for applications in the chemical industry and energy sector. An attractive system for high efficiency photoelectrochemical water splitting consists of two semiconductors connected in series and each coupled to a metal electrocatalyst. An efficiency analysis of these dual semiconductor systems demonstrates that the optimal band gap for the bottom absorber is between 1-1.4 eV [1]. Silicon has an optimal band gap of 1.1 eV, but it is chemically unstable in relevant water splitting conditions [2]. To overcome these stability concerns, metal-insulator-semiconductor (MIS) systems have emerged as a promising strategy. In MIS photocatalysts, the metal layer serves as the catalyst and the insulator serves as a protective layer. Besides improving stability, these MIS systems can be engineered to maximize the light absorption and the generated photovoltage. Previous research has shown that the insulator layer thickness is an important parameter that can be tuned to optimize the photovoltage [3,4]. The objective of this work is to further elucidate the mechanism by which the nanoscopic control of the insulator thickness enhances the performance of MIS photoelectrocatalysts and develop design principles to achieve even higher efficiencies. We will demonstrate our results by using an Ir/HfO_2

experimental results to a theoretical model. This approach is general to other insulators and provides insights to identify favorable insulator material properties and further optimize the photovoltage in these MIS systems for water splitting applications.

References

1. Seitz, L.C., Chen, Z., Forman, A.J., Pinaud, B.A., Benck, J.D., Jaramillo, T.F. *ChemSusChem* 7, 1372, (2014).
2. Hu, S., Lewis, N.S., Ager, J.W., Yang, J., McKone, J.R., Strandwitz, N.C. *Phys. Chem. C* 119, 24201, (2015).
3. Quinn, J.P., Hemmerling, J.R., Linic, S. *ACS Catal.* 8, 8545, (2018).
4. Digdaya, I.A., Trzesniewski, B.J., Adhyaksa, G.W., Garnett, E.C., Smith, W.A., *Phys. Chem C* 122, 5462, (2018).

573689 Directed Self-Assembly of Block Copolymers for Well-Defined Nanostructured Electrocatalysts

Periodic nanostructures with long-range order and low-level structure defects are hallmark achievements of Directed Self-Assembly (DSA) of block copolymers for lithographic patterning. This bottom-up, macromolecular nanomanufacturing platform has been adopted by numerous industrial companies and consortiums for developing the next-generation microelectronics with smaller feature sizes that enable greater data processing speeds and storage. However, the DSA platform has largely been confined to the patterning of semiconductor materials and little attention has been given to the potential it offers for electrochemical applications. In particular, the DSA platform for fabricating well-defined electrocatalysts with high fidelity is intriguing as model nanoscale electrocatalyst studies often do not feature uniform spacing and a single morphology when processed as thin electrodes for electrocatalytic activity studies (e.g., in a rotating disk electrode setup). The lack of well-defined electrocatalyst structures hinder structure-property relationships that definitively reveal how nanoscale attributes govern electrocatalytic activity.

This talk will present two different DSA nanomanufacturing platforms to access platinum, palladium, and gold nanowires and dots (and alloys of these metals) from perpendicular lamellae and cylinder block copolymer templates on electron conducting glassy carbon substrates and interdigitated electrode substrates. The two nanomanufacturing methods consisted of: i.) incipient wetness impregnation of self-assembled poly(styrene-block-2-vinyl-methyl pyridinium iodide/pyridine) with metal salts followed by reduction and ii.) self-assembled block copolymers were one polymer block is etched out, physical vapor deposition of electrocatalyst metals occurs followed by liftoff of the porous polymer mask. X-ray photoelectron spectroscopy and electron microscopy were used to characterize the morphology and chemical composition of the extended surface electrocatalysts. Catalytic activity for model reactions (e.g., oxygen reduction and water oxidation) were substantiated with the well-defined nanostructure materials. Connections between morphology to electrocatalytic activity will be disseminated.

563131 The Novel Electrocatalyst/Semiconductor Photoelectrodes for Efficient Photoelectrochemical Water Splitting

Water splitting by solar-driven photoelectrochemical (PEC) reactions has been broadly recognized as one of the most promising chemical method for sustainable energy conversion and storage systems. Effective integration of semiconductor and electrocatalyst is critical for fabricating efficient PEC devices. However, this is a big challenge, because the introduction of interfacial recombination centers on the electrocatalyst/semiconductor junction result in charge recombination. Moreover, considering the advantages of electrocatalysts with ultrathin morphology, the rational design of electrocatalyst/semiconductor architecture to improve the hydrogen evolution reaction (HER) or oxygen evolution reaction (OER) kinetics without sacrificing light harvesting is of significance to achieve efficient PEC system. Recently, we found that amorphous materials synthesized by a facile electrodeposition method can serve as highly efficient electrocatalysts for efficient photoelectrochemical water splitting.

Transition metal chalcogenides (TMC) are considered as the promising HER electrocatalysts. The electrodeposition method can produce various TMC HER catalysts such as $CoSe_2$, $Ni_{0.85}Se$, $NiSe_2$ and $CoSe_x$ etc. Also we realized the rational design of a novel 3D p-Si/$NiCoSe_x$ core/shell nanopillar (NP) array photocathode by uniform photo-assisted electrodeposition of $NiCoSe_x$ electrocatalyst on bamboo shoots-like Si NP array backbones. We demonstrate its enhanced PEC performance with a photocurrent density of -37.5 mA/cm^2 at 0 V (vs. RHE) under simulated 100 mW/cm^2 (1 Sun) with an AM 1.5 G filter, which is the highest value reported for p-type Si photocathodes. The synergic effects of the excellent light harvesting of Si NP array core and the good optical transparency, as well as excellent electrocatalytic activity of $NiCoSe_x$ shell boost the production and utilization of photo-generated electrons [3-5].

Furthermore, metal hydroxides are excellent electracatalysts for water splitting. It was found that amorphous cobalt-iron hydroxide (CoFe-H) nanosheets by the electrodeposition can be simply integrated with $BiVO_4$ semiconductor to construct CoFe-H/$BiVO_4$ photoanodes that exhibit a significantly enhanced photocurrent density of 2.48 mA/cm^2 (at 1.23 V vs. RHE) and a much lower onset potential of 0.23 V (vs. RHE) for PEC-OER. Careful electrochemical and optical studies reveal that the improved OER kinetics and high-quality interface at the CoFe-H/$BiVO_4$ junction, as well as the excellent optical transparency of CoFe-H nanosheets, contribute to the high PEC performance.

Our work proves that the integrated electrocatalyst/semiconductor photoelectrodes are active for high performance solar-driven photoelectrochemical water splitting. We think high-quality electrocatalyst/semiconductor interface and excellent optical transparency of electrocatalysts result in the high PEC performance.

573848 Biomineralized Metal Chalcogenide Quantum Dot/Rgo Photocatalysts for Water Splitting

Metal chalcogenide quantum dots (QDs) are promising materials for light harvesting in photocatlytic water splitting. However, these materials are typically synthesized through high temperature, multi-step processes that utilize organic solvents and expensive precursors. These factors increase the complexity and economic and environmental costs of manufacturing scale-up, especially when we consider the desired scale of photocatalytic hydrogen production. In contrast, biomineralization, the process by which biological systems produce structural nanomaterials, occurs under ambient conditions in aqueous media utilizing the available precursors. We have developed routes to adapt these biological processes to the direct, scalable, size-controlled synthesis of metal chalcogenide quantum dots in numerous forms; from single component materials, to alloys, and core-shell heterostructures. We have also developed an analogous route to the synthesis of reduced graphene oxide (rGO) utilizing biologically generate reductant. In combination, these routes enable biofabrication of highly active QD/rGO photocatalysts. This presentation will cover the origin of size control in this synthesis process, the optimization of the biomineralization processes and QD-rGO linker molecules, and demonstrate the activity of the resulting photocatalyst for solar hydrogen generation.

562938 Multi-Scale Ordering in Highly Stretchable Polymer Semiconducting Films

Stretchable semiconducting polymers have been developed as a key component to enable skin-like wearable electronics, but their electrical performance must be improved to enable more advanced functionalities. Here, we report a solution processing approach that can achieve multi-scale ordering and alignment of conjugated polymers in stretchable semiconductors to substantially improve their charge carrier mobility. Using solution-shearing with a patterned microtrench coating blade, the macroscale alignment of conjugated-polymer nanostructures was achieved along the charge transport direction. In conjunction, the nanoscale spatial confinement aligns chain conformation and promotes short-range p–p ordering, which significantly reduce the energetic barrier for charge carrier transport. As a result, the mobilities of stretchable conjugated-polymer films have been enhanced by up to three-fold and maintained under stain up to 100%. This method can also serve as the basis for large-area manufacturing of stretchable semiconducting films, as demonstrated by the roll-to-roll coating of meter-scale films.

575127 Impact of Helical Chain Shape in Sequence-Defined Polymers on Polypeptoid Block Copolymer Self-Assembly

Controlling the self-assembly of block copolymers with variable chain shape and stiffness is important for driving the self-assembly of functional materials containing nonideal chains. The polymer helix is a particularly important functional motif in which the scaling relationships between local chain stiffness and space-filling properties impact self-assembly. Polypeptoids, a class of sequence-defined peptidomimetic polymers with controlled helical secondary structure, were used to systematically investigate the impact of helical chain shape on block copolymer self-assembly in a series of poly(*n*-butyl acrylate)-*b*-polypeptoid block copolymers. By leveraging sequence control, the precise location of the helix-forming chain section relative to the block junction was controlled. Small-angle X-ray scattering (SAXS) revealed that chain stretching of the helix near the block junction is disfavored, while helical segments at the center of cylindrical domains contribute to unfavorable packing interactions. Finally, temperature-dependent SAXS shows that helix-containing diblock copolymers disorder at lower temperatures than the equivalent unstructured diblock copolymers. These results emphasize that current descriptions of rod/coil interactions and conformational asymmetry for coil polymers do not adequately address the behavior of chain secondary structures in controlling self-assembly.

575536 Combining Polymer Synthesis with Self-Assembly of Block Copolymers

Self-assembly of polymers has emerged as an important and promising technique for nanopatterning due to its advantages of low cost and high versatility. The conventional strategy requires two separate steps including (1) chemical synthesis and (2) physical assembly, where the resulting morphology is governed by the underlying polymer chemistry. Altering the nanostructures of block copolymer (BCP) films from its thermodynamic equilibrium state is challenging, due to our very limited control over chemistry during or after self-assembly. Integrating polymer synthesis within a self-assembled film would enable exact on-demand control over the resulting morphology, functionality, and properties in different regions.

In this talk, we will demonstrate a new approach for performing polymerization reactions within block copolymer thin films. The vapor phase monomer is first introduced into the block copolymer film, which can be subsequently converted to polymers through the photopolymerization. The newly synthesized homopolymers blend with BCP films and alter the nanostructures by changing the underlying polymer chemistry and composition. Identical photopolymerization conditions with lower UV intensity causes more expansion of selective domains due to the addition of higher molecular weight homopolymers, leading to highly asymmetric nanopatterns. As these altered nanostructures are locally near equilibrium, this method can be adapted and combined with common annealing techniques to improve order. This strategy also enables a stable phase of toroidal micelles by polymerizing a monomer that has distinct chemistry compared to the block copolymer chemical composition. By successfully combining polymer chemistry with self-assembly, this in-film polymerization method provides an exciting platform for on-demand synthesis as well as opening up a new area for radical polymerizations of monomers within geometrically confined, swollen films.

573676 3D Analysis of Nanostructured Polyamide Membranes Using Quantitative Electron Tomography

State-of-the-art reverse osmosis, nanofiltration, and desalination membranes are constructed as a thin film composite: they consist of crosslinked polyamide selective layer with an extremely thin thickness (i.e., <100 nm). These thin film membranes are generally prepared via interfacial polymerization between aromatic or aliphatic diamine and aromatic acyl chloride. In general, their performance of polyamide membranes is governed by their morphology. However, the interfacial polymerization process leads to highly irregular and complex crumpled nanostructures that make the quantitative measurement of their physiochemical properties extremely difficult. Here, we demonstrated 3D electron tomographic imaging and quantitative morphometry of polyamide films to understand their synthesis-structure-property relationship. We employed a series of interfacially polymerized membranes with varying monomers concentrations to determine how complex nanoscale morphology emerge from different synthesis conditions. We reconstructed the crumpled morphology in 3D with a voxel of 0.68 nm using low-dose electron tomography. Using our imaging-quantitative morphometry platform, we observed that the effective thickness films increased and the film structure became more crumpled with decreasing the concentration of an acyl chloride monomer and/or increasing that of a diamine monomer. We observed multimodal thicknesses in each reaction condition. This observation supported a crumpling mechanism in which uniform pieces of polyamide film are formed in the early stage of polymerization and then a local collapsing process serves as the basis for the emergence of 3D crumple structures. Furthermore, various parameters governing film transport properties, such as surface-to-volume ratio and mass-per-area, were measured directly from the reconstructed membrane structures. We conducted local curvature mapping and metal ion adsorption tests to understand how reaction condition affects the microenvironment of the crumples. We anticipate that this imaging-morphometry platform can be applicable to other nanoscale soft materials and provides engineering strategies based directly on synthesis-morphology-function relationships.

569935 Solvent Vapor Annealing to Control Polymer Crystal Morphology

Degree of crystallinity, crystal morphology, and crystallite size affect many physical properties of semi-crystalline polyesters, including mechanical strength, optical clarity, and biodegradability. Inspired by the literature related to solvent vapor annealing (SVA) in block copolymer films and solvent-induced crystallization in semi-crystalline polymers, we are studying how SVA treatments impact crystal morphology. Using poly(e-caprolactone) as a model polymer for our studies, we have conducted in-situ grazing incidence wide-angle X-ray scattering experiments to determine when/if polymer crystals dissolve during solvent uptake and at what solvent concentrations recrystallization occurs during solvent removal. Additionally, we have examined polymer morphology across multiple length scales using a combination of optical microscopy, profilometry, and atomic force microscopy.

568942 Cation Conduction in Nanostructured Side-Chain Ionomers

Solvent-free polymer electrolytes have long been sought after for use in rechargeable batteries. The ideal electrolyte would have high ionic conductivity, a high active ion transference number, a wide electrochemical stability window, and chemical stability with the electrodes of interest. Poly(ethylene oxide)-based electrolytes have historically received the most research attention despite cation conduction rates being highly dependent on chain segmental motion. Here we report on non-polyether, cation conducting, polyanion-based electrolytes. Single-ion conducting, side-chain polymer electrolytes are investigated wherein lithium conduction occurs in ionic domains or aggregates. The effect of polymer design and tethered anion chemistry on phase segregation and lithium transport will be discussed.

574443 Poly(ionic liquid) Pentablock Terpolymer Electrolytes in Solid-State Lithium Ion Batteries

New solid polymer electrolytes (SPEs) that comprehensively possess the preferred properties of high ion conductivity, high electrochemical stability, robust mechanical properties, flexibility, and good film forming properties, can improve the safety and overall charge capacity and lifetime of lithium ion batteries. Although diblock copolymers have been explored as SPEs and can potentially provide the orthogonal properties of high ion conduction and high mechanical strength simultaneously in a solid-state material, there are limitations to lithium ion conducting diblock copolymers, such as a limited set of morphologies, and a lack of means to achieve simultaneous combination of more than two properties (*e.g.*, conductivity, strength, flexibility). Compared to diblock copolymers, multiblock polymers involve more than two polymer chemistries and therefore enable the possibility to conjoin many of the desired properties all into a single platform. In our previous work, a lithium ion conducting poly(ionic liquid) pentablock terpolymer was explored as an SPE in lithium ion batteries and showed promising room temperature battery performance (charge-discharge cycles).In this study, a fundamental understanding of the relationship between electrochemical/chemical/mechanical properties and battery performance was investigated. A combination of *in situ* and *ex situ* experiments, such as battery performance, electrochemical impedance, stripping and platting, infrared spectroscopy, electron microscopy, x-ray photoelectron spectroscopy, and dynamic mechanical analysis, were conducted and will be presented.

579269 Self-Assembly and Ion Transport of Hybrid Organic-Inorganic Diblock Copolymers with Salt

We examine the phase behavior of a hybrid organic-inorganic diblock copolymer/salt mixtures. The experimental system comprises poly(ethylene oxide)-*block*-polyhedral oligomeric silsesquioxane (PEO-POSS) mixed with a lithium bis(trifluoromethanesulfonyl) imide (LiTFSI) salt. While the diblock copolymers without salt exhibit classical order-to-disorder transition behavior with increasing temperature, the PEO-POSS/salt mixtures exhibit disorder-to-order transitions with increasing temperature. We posit that thermodynamic POSS and salt ions are responsible for this unusual behavior. The relationship between the self-assembled morphology and ion transport will be discussed. In particular, these electrolytes exhibit a relatively high limiting current.

575125 Self-Assembled ssDNA Nanotubes for Targeting Cancer (Invited Speaker)

Self-assembly of biological molecules is an attractive method for engineering supramolecular biomaterials for different applications. In my group we focus on the design and characterization of amphiphilic molecules, such as ssDNA-amphiphiles, that have the tendency to self-assemble spontaneously in water. We have shown that the type of molecular spacer linking a single-stranded DNA (ssDNA) headgroup to a lipid-like tail affects the assembly of the ssDNA-amphiphiles, producing spherical micelles, bilayer nanotapes (flat or twisted) and nanotubes. I will discuss how we use ssDNA nanotubes to target cancer, *in vitro* and *in vivo*.

565386 Self-Assembly of Hydrophobically-Patterned Charged Polypeptides for Drug Delivery

Polyelectrolyte complexation is the result of the interaction of oppositely charged macromolecules in solution. Under precise conditions, these complexes can form liquid-liquid phase separated states called complex coacervates. The complex coacervate phase has a very low interfacial tension with water and therefore can be used as an encapsulant in the food and pharmaceutical industries. Here we design hydrophobically-patterned ionic polypeptides in order to systematically study the effect of hydrophobicity on complex behavior and to assess the ability of these hydrophilic materials to encapsulate hydrophobic drugs. Two generations of oppositely charged peptides are designed, containing non-charged amino acids of increasing hydrophobicity: glycine, alanine, and leucine; while lysine and glutamic acid are used as the charged amino acids for the polycation and polyanion sequences, respectively. Previous work with ionic polypeptides has determined that chirality is an important factor that controls the phase behavior of polypeptide complexes, where alternating patterns of L and D amino acids leads to complex coacervate formation, while homochiral sequences form solid precipitates. Therefore, here we use an alternating sequence of D and L-chiral amino acids in order to promote liquid coacervate formation. The first generation designs consist of D-charged residues and L-hydrophobic residues, while the second generation peptides, also have alternating chiral patterns, but a higher charge density. Peptide sequences are characterized using mass spectroscopy (MALDI-TOF), H NMR and circular dichroism (CD). The polypeptide based polyelectrolyte complexes are characterized using turbidity measurements to evaluate the extent of complex formation as well as the stability of the polyelectrolyte complexes against salt (NaCl) and temperature. In addition, optical microscopy and infrared spectroscopy are used to characterize complex phase behavior and secondary structure. Our results indicate that increasing hydrophobicity improves the stability of the complexes and increases the amount of complex formed with increasing temperature. Moreover, the density and amount of the complex phase are found to increase with hydrophobicity. We use a hydrophobic model dye to assess the encapsulation behavior of the complexes showing that encapsulation efficiency of the complexes increases as the hydrophobicity increases. This work provides insight for the potential application of these complexes in drug delivery containing both charged biological therapeutics and hydrophobic drugs.

588760 A Multifunctional DNA-origami Platform to Sense Cellular Interactions In Situ

The plasma membrane is the primary communicating interface between the cell and the extracellular environment. Thus, the ability to engineer new functions into the cell membrane to externally monitor cellular interactions is of great significance. Previous studies have focused on engineering cell surface proteins or incorporating synthetic protein constructs into the cell membrane, which are often highly challenging. Therefore, designing a robust approach to incorporate multifunctional nanodevices with diverse structural and dynamic functions into the cell membrane can enable engineering of the cell surface as a functional material. Furthermore, none of the available techniques provide in-situ monitoring of cell-biomarker interaction. Thus, we aim to engineer DNA-based detection devices on the cell membrane to detect the spatiotemporal distribution of cancer biomarkers in situ. Our initial focus is to detect circulating tumor DNA (ctDNA), mutated genes that are released by dead cancer cells, and protein thrombin in-situ with fluorescence-based reporting. We further aim to combine ctDNA and thrombin detection with other functions like sensing pH, other proteins, and physical forces at the cell surface with the aim of spatially and temporally profiling cellular interactions in tumor microenvironments. Ultimately, this project could lead to the development of a multi-functional in-situ detection method, capable of elucidating on the spatiotemporal distribution information of cell-environment interaction.

574850 Peptoid Microsphere Coatings: The Effects of Helicity, Temperature, pH, and Ionic Strength

Peptoids are peptidomimetic oligomers that predominantly harness similarities to peptides for biomimetic functionality. They have potential for use in biomedical applications and biosensors due to resistance to proteolytic degradation and low immunogenicity. The incorporation of chiral, aromatic side chains in the peptoid sequence allows for the formation of distinct secondary structures and self-assembly into supramolecular assemblies, including microspheres. Peptoid microspheres can be coated onto substrates for potential use in biosensor technologies, tissue engineering platforms, and drug-delivery systems. In order to be useful for these applications, the peptoid coatings must be robust under physiological conditions. In this study we report the effects of various conditions on the peptoid microsphere coatings, including (i) helicity, (ii) temperature (iii) pH, and (iv) ionic strength. These studies show that microsphere size decreases with increasing peptoid helicity and the positively charged side chains are positioned on the outside of the microspheres. The peptoid microsphere coatings are robust under physiological conditions but degrade in acidic conditions ($pH < 7$) and at low ionic strengths ($< 150\ \mu M$).

562677 Cationic, Anionic, and Amphoteric Dual pH/Temperature-Responsive Degradable Microgels Via Self-Assembly of Functionalized Oligomeric Precursor Polymers

Background:

Stimulus-responsive or 'smart' materials are increasingly attracting interest in biotechnology and medicine, especially in the context of drug-delivery systems, given their potential to rapidly switch properties (e.g. size, hydrophobicity, charge, etc.) when exposed to small changes in their environment (i.e. pH and temperature) [1]. These changes are reversible, allowing the system to return to its initial state when the environmental trigger is removed. In particular, "smart" microgels that can change their properties in response to physical (temperature, light, electric and magnetic fields) or chemical (pH, ionic strength, and the presence of chemical or biological compounds) stimuli have been identified to have various applications including sensors [2][3], optics [4] and loading/release applications [5][6]. Temperature-responsive poly(N-isopropylacrylamide) (PNIPAM) microgels are the best-known example [7][8]. PNIPAM microgels have a volume phase transition temperature (VPTT), a direct result of the lower critical solution temperature (LCST) of the PNIPAM polymer (~32°C) [9]; consequently, PNIPAM microgels are swollen at room temperature but, when heated above the VPTT, they discontinuously shrink, inducing tunable changes in diameter, hydrophobicity, pore size and surface charge as a function of temperature.

Conventionally, PNIPAM microgels are fabricated using free-radical precipitation polymerization, with a free radical initiator, a monofunctional monomer (giving the linear segments), a bifunctional monomer (giving crosslinks) [10]. By copolymerizing acidic or basic comonomers, pH-responsiveness can be introduced to create dual responsive microgel systems useful for ion exchange, drug delivery, environmental sorption, and other applications. However, to use such microgels for biomedical applications, degradability remains an issue. Most microgel preparations use a non-degradable crosslinker; even if the crosslinker is degradable, the molecular weights of the degradation products are often not controllable and thus may not be cleared following gel degradation.

In response, we have developed a novel approach to overcome this limitation by creating degradable thermosensitive microgels by the controlled precipitation of well-defined hydrazide and aldehyde-functionalized PNIPAM oligomers [11][12]. This thermally driven self-assembly approach enables the formation of a hydrolytically-labile hydrazone cross-linked network that can break down over time into oligomeric by-products of molecular weights below the renal filtration limit. These self-assembled PNIPAM microgels are degradable, monodisperse, homogeneously crosslinked, and stable over months, while retaining their responsiveness to temperature due to the presence of PNIPAM.

Objectives:

By copolymerizing charged monomers into the precursor polymers used in self-assembly, pH-responsive swelling behaviors can be achieved that may are relevant for a number of applications ranging from sensors to drug delivery. The primary and foremost objective of this work is to synthesize and characterize single-charged and amphoteric PNIPAM microgels prepared using the self-assembly method. Next, the internal structure of these charged microgels are analyzed using small-angle and ultra-small angle neutron scattering experiments. Prior to *in vitro* drug release studies, the microgels were exposed to C2C12 myoblast cells to evaluate their cytocompatibility. Lastly, moderately hydrophobic drugs (i.e. naproxen sodium) were loaded into these microgels using two methods, and the encapsulation efficiencies and drug release kinetics were analyzed.

Methodology:

Hydrazide and aldehyde-functionalized PNIPAM polymers were synthesized via free radical copolymerization of NIPAM in ethanol using thioglycolic acid as the chain transfer agent and 2,2-

azobisisobutyric acid dimethyl ester as the initiator. Cationic and anionic charges are incorporated into the self-assembled PNIPAM microgels by functionalizing either functionalized polymer with anionic (acrylic acid) and cationic (N,N-dimethylaminoethyl methacrylate) comonomers. Titration, NMR and GPC data was collected to characterize each precursor polymer. The hydrazide-functionalized PNIPAM polymer is dissolved and heated to 5+LCST for 2-3 minutes, after which the aldehyde-functionalized PNIPAM polymer is added dropwise at 20 mass%, and the self-assembly is complete after ~15 minutes. Incorporation of pH-ionizable monomers into the precursor polymers, and subsequent self-assembly of those polymers will result in dual pH/temperature-responsive microgels. Charged oligomers are added in different combinations and sequences, enabling precise tuning of the charge density and charge distribution based on the net charge and sequence of assembly of the precursor polymers used (e.g. amphoteric microgels with different charge profiles can be fabricated by mixing cationic and anionic-functionalized precursor polymers at different stages during the self-assembly process). The optimal conditions associated with the production of these pH-responsive PNIPAM microgels were explored and evaluated in terms of size (dynamic light scattering), charge (electrophoretic mobility), internal morphology (small angle neutron scattering, SANS) and degradation (light scattering). Cell compatibility was evaluated using fluorescently labeled C2C12 myoblast cells and performing an MTS assay. Moderately hydrophobic drugs (i.e. naproxen sodium) were co-self-assembled with the polymers to attempt to optimize drug loading by avoiding partitioning-based effects that typically limit drug loading in conventional microgels. Drug release experiments were conducted using membrane devices, with the amount of drug released assessed using high performance liquid chromatography.

Results:

Self-assembled single-charged PNIPAM microgels exhibit similar pH and temperature-responsive particle sizes and zeta potentials as conventional microgels, with the added advantage of degradability into polymers with known molecular weights below the kidney clearance limit. Amphoteric microgels can also be successfully fabricated by mixing cationic and anionic-functionalized precursor polymers during the self-assembly process, reproducing the high pH/low pH parabolic swelling response observed in conventional amphoteric microgels. The internal structure was analyzed using SANS and USANS experimental data and compared to the homogenous internal structure of the uncharged self-assembled PNIPAM microgels. No major structural differences can be observed despite the introduction of charge into the system. Cytotoxicity studies using an MTS assay and C2C12 mouse myoblast cells indicated no cytotoxic effects are observed for the charged microgels or the precursor polymer degradation products (>80% viability). The common drug loading method for conventional precipitation based PNIPAM microgels is through passive diffusion of the drug into the polymer network. This method of drug loading is slow (>24 hours) and typically results in low encapsulation efficiencies due to the incompatibility between the hydrophobic drug and hydrophilic crosslinked networks of the microgels. In contrast, the radical-free self-assembly method also allows for in situ drug loading by directly co-assembling the drug with the prepolymers, facilitating between 3-5 times higher drug loading that is completed within <15 minutes (concurrent with particle fabrication) for moderately hydrophobic drugs (e.g. dexamethasone, naproxen sodium), which preferentially partition inside the hydrophobic nanoaggregate formed during the self-assembly process. In addition, due to the resulting different distribution of drug inside the microgels following *in situ* versus passive release, no significant burst release was observed for the naproxen-loaded microgels and near zero-order drug release continued over the course of two weeks, significantly longer than achieved with conventional passive drug loading.

Significance:

The significance of this self-assembly approach relative to conventional microgel synthesis is based on (1) the potential for degradation via hydrolysis and (2) the ability to form particles by simple sequential mixing within minutes. Introduction of pH-sensitivity to the already temperature-responsive self-assembled PNIPAM microgels allows for multiple stimuli-responsive properties. Coupling the precise dual responsive swelling responses achievable with the degradability of the hydrazone crosslinks, self-assembled charged PNIPAM microgels offer potential for improved performance in drug delivery applications demanding dual pH/temperature specific delivery (e.g. targeting infection sites or cancer).

References:

[1] I. Y. Galaev and B. Mattiasson, "'Smart' polymers and what they could do in biotechnology and medicine," *Trends Biotechnol.*, vol. 17, no. 8, pp. 335–340, Aug. 1999.

[2] D. Wang, T. Liu, J. Yin, and S. Liu, "Stimuli-Responsive Fluorescent Poly(N-isopropylacrylamide) Microgels Labeled with Phenylboronic Acid Moieties as Multifunctional Ratiometric Probes for Glucose and Temperatures," *Macromolecules*, vol. 44, no. 7, pp. 2282–2290, Apr. 2011.

[3] J. Hu and S. Liu, "Responsive Polymers for Detection and Sensing Applications: Current Status and Future Developments," *Macromolecules*, vol. 43, no. 20, pp. 8315–8330, Oct. 2010.

[4] Y. Gao, X. Li, and M. J. Serpe, "Stimuli-responsive microgel-based etalons for optical sensing," *RSC Adv.*, vol. 5, no. 55, pp. 44074–44087, 2015.

[5] Y.-J. Pan *et al.*, "Redox/pH dual stimuli-responsive biodegradable nanohydrogels with varying responses to dithiothreitol and glutathione for controlled drug release," *Biomaterials*, vol. 33, no. 27, pp. 6570–6579, Sep. 2012.

[6] D. Klinger and K. Landfester, "Stimuli-responsive microgels for the loading and release of functional compounds: Fundamental concepts and applications," *Polymer*, vol. 53, no. 23, pp. 5209–5231, Oct. 2012.

[7] J. McMasters, S. Poh, J. B. Lin, and A. Panitch, "Delivery of anti-inflammatory peptides from hollow PEGylated poly(NIPAM) nanoparticles reduces inflammation in an ex vivo osteoarthritis model," *J. Controlled Release*, vol. 258, pp. 161–170, Jul. 2017.

[8] F. A. Plamper and W. Richtering, "Functional Microgels and Microgel Systems," *Acc. Chem. Res.*, vol. 50, no. 2, pp. 131–140, Feb. 2017.

[9] Y. Guan and Y. Zhang, "PNIPAM microgels for biomedical applications: from dispersed particles to 3D assemblies," *Soft Matter*, vol. 7, no. 14, pp. 6375–6384, Jul. 2011.

[10] R. H. Pelton and P. Chibante, "Preparation of aqueous latices with N-isopropylacrylamide," *Colloids Surf.*, vol. 20, no. 3, pp. 247–256, Oct. 1986.

[11] D. Sivakumaran, E. Mueller, and T. Hoare, "Temperature-Induced Assembly of Monodisperse, Covalently Cross-Linked, and Degradable Poly(N-isopropylacrylamide) Microgels Based on Oligomeric Precursors," *Langmuir*, vol. 31, no. 21, pp. 5767–5778, Jun. 2015.

[12] E. Mueller *et al.*, "Dynamically Cross-Linked Self-Assembled Thermoresponsive Microgels with Homogeneous Internal Structures," *Langmuir*, vol. 34, no. 4, pp. 1601–1612, Jan. 2018.

576407 Calcium-Phosphate Supraparticle Assembly Is Influenced By Both Intermolecular Forces and Fluid Dynamics

The formation of hierarchical composite materials is ubiquitous in nature. Colloidal interactions between nanostructures can lead to complex micro- and macrostructures with internal pore networks, enhanced physical properties, and order on many scales. A mixture of calcium, phosphate, and protein leads to self-assembly of nanocomposites. However, these samples are often polydisperse and loosely ordered. We previously reported that an end-over-end rotation of such mixtures with contraction of the air-water interface can form microparticles with a regular structure. So-called supraparticles are composed of calcium-phosphate and a leucine zipper binding protein, which allows for the specific immobilization of proteins with the complementary binding domain.

Building off this work, we have uncovered the parameter space for the formation of uniform particles. Using microscopy and a specific binding assay, we sought to identify conditions for supraparticle formation to better understand the self-assembly process. The reaction medium composition was varied via component concentrations, ionic strength, and pH in a 3x3x3 Design of Experiment matrix. Once the conditions with supraparticle morphology were uncovered, the effects of additives such as DMSO and glycerol were tested, demonstrating the influence of viscosity and hydrogen bonding on the formation of supraparticles. With the ideal composition of reagents in the solution for particle formation, we elucidated the effects of tube geometry and flow regime on the output. Scalability of the synthesis procedure was considered using dimensional analysis. The supraparticles created here can be applied to immobilized enzyme systems, where a scalable material with predictable binding behavior would be preferred for a reactor.

566756 Interaction of Amphiphilic Coumarin with DPPC/Dpps Using DSC and MD Simulations: Effects of Concentration and Chain Length

Amphiphilic molecules have better cell penetration properties. Interactions between novel amphiphilic amino methyl coumarin with DPPC/DPPS liposomes were investigated by combining experimental (zeta potential, fluorescence spectroscopy, DSC) and theoretical MD simulations studies. Alkyl chains comprising of 5, 9 and 12 carbon atoms were conjugated to amino methyl coumarin via single step process. Our results indicate that C5 and C12 associate with lipid membrane via flip-flop mechanism. At higher concentrations, C5 aggregate resulting in continuous insertion and ejection from the outer leaflet of the lipid bilayer causing thinning of the membrane. On the contrary, C12 lacked the ejection-reinsertion behavior suggesting they were associated with the inner leaflet of the bilayer and flip-flop occurred between the outer and inner leaflets resulting in membrane destabilization. This was further confirmed by the DSC studies that showed disappreance of phase separation and a well-mixed lipid system in the bilayer with subsequent annealing. The presence of broader transition peak also indicated increased melting cooperativity or associated with increased number of lipid molecules. The efficacy of C12 coumarin for potential biomedical applications were further demonstrated via preliminary cellular uptake, toxicity and photodynamic activity in cancer cells.

578229 Magnetic Particle Imaging for In Vivo Quantitative Tracking of Nanoparticle Drug/Gene Carriers

Magnetic Particle Imaging (MPI) is a new molecular imaging technology capable of unambiguous and quantitative tomographic imaging of the distribution of superparamagnetic nanoparticle tracers in vivo. While the term MPI may be confused with that for Magnetic Resonance Imaging (MRI), the two are distinct and rely on different physics. An MPI image cannot be generated using an MRI machine and vice versa. In MPI, a tomographic image of the distribution of superparamagnetic nanoparticles is constructed by scanning a so-called field free region (FFR) through the domain of interest. Outside the FFR there is a quasi-static bias field strong enough to saturate the magnetic moments of the nanoparticles. But inside the FFR the dipole moments of the nanoparticles are able to respond to a superimposed alternating excitation field. The signal used to construct an image in MPI arises due to the non-linear dynamic magnetization response of the nanoparticle dipole moments to the excitation field inside the FFR. At the field amplitudes and frequencies used in MPI there is no appreciable attenuation in field or signal strength in tissue. Further, while there are magnetic species in the body (e.g., ferritin), they do not contribute an appreciable signal for MPI, allowing for unambiguous imaging of the distribution of one of the superparamagnetic nanoparticle tracers.

Superparamagnetic nanoparticles have been widely studied as components in drug and gene nanoparticle carriers. This is often motivated by the excellent biocompatibility of iron oxide nanoparticles; the ability to actuate motion and heat release through external applications of magnetic field gradients or alternating magnetic fields; and the ability to image their distribution through magnetic resonance imaging. However, superparamagnetic iron oxide nanoparticles generate a hypo-intense T2 weighted signal that is easy to confuse with air/blood artifacts and that is difficult to quantify. Despite this, many studies attempt to visualize and even quantify the distribution of superparamagnetic iron oxide nanoparticles using magnetic resonance imaging.

In this talk I will explain the physics of image generation in MPI, discuss work to understand how imaging performance relates to physical and magnetic properties of the nanoparticles, and discuss work (my own and that of others) using MPI to quantify biodistribution of iron oxide nanoparticles in vivo, in the context of nanoparticle drug carriers and in other contexts.

575858 pH-Responsive, Core-Shell Magnetite-Silver Nanoparticles for the Guided Transport and Delivery of Nucleotide Cargoes: An Avenue for Highly-Targeted Gene Therapies

Over the past few years, gene therapies have attracted much attention for the development of therapies for various conditions including cancer, neurodegenerative diseases, protein deficiencies and autoimmune disorders. Despite the benefits of this approach, several challenges are yet to be solved to eventually reach clinical implementation. Some of these challenges include low transfection rates, limited stability under physiological conditions and poor specificity towards the target cells. An avenue to overcome such issues is to deliver the therapies with the aid of potent cell-penetrating vectors, which include viral and non-viral vehicles. The use of viral vectors has been criticized due to issues concerning genome integration, a potential for oncogenicity and induced mutagenesis in the host cell1. Alternatively, non-viral vectors, such as nanostructured materials, have been successfully tested in drug and gene delivery without the concerns of the viral ones 2. This has been attributed to their high surface area to volume ratio, ease of synthesis, and the possibility for surface modification to incorporate various surface chemistries and thereby a tunable multifunctional response. This has led to an increasing number of applications where the delivered molecules have shown a considerable increase in bioavailability in both in vitro and in vivo applications3. Among the most successful nanostructured vehicles, magnetite nanoparticles stand out due to their superior biocompatibility and the possibility for manipulation with the aid of magnetic fields.

Here, we propose to develop a nanostructured core-shell cell-penetrating vehicle composed of magnetite at the core and surrounded by a silver shell. A subsequent superficial conjugation of a pH-responsive polymer was proposed to assure that the vehicle is capable of carrying and releasing circular DNA. Accordingly, the magnetite core was synthesized by co-precipitation of 500mM ferric chloride and 250mM ferrous chloride (molar ratio 2:1) in the presence of a 5M NaOH aqueous solution at 90°C under mechanical agitation at 300 rpm. This was followed by deposition of the silver from a 1 mM silver nitrate solution in a reducing honey solution 20% (w/v). During the process, pH was maintained above 4 to prevent undesirable degradation reactions. Silver on the surface was subsequently chlorinated with the aid of a 1M HCl solution. The prepared core-shell system exhibited colloidal stability such that agglomerate formation and thereby precipitation was avoided. A Hofmann elimination reaction was further conducted on the system to conjugate the pH-responsive polymer poly(2-dimethylamino) ethyl methacrylate (pDMAEMA) to the chlorine atoms on the surface. The conjugation reaction proceeded at 50°C under continuous mechanical agitation at 500 rpm. Besides the role as DNA carrier, the pDMAEMA was conjugated to attempt to avoid endosome formation, which is a cell internalization mechanism that traps exogenous material and consequently might lead to a significant reduction in transfection rate4. Finally, a 10kb circular DNA vector was loaded on the modified nanoparticles at pH values in the range of 6 to 8.5 to estimate the loading efficiency, which was defined as the DNA trapped by the system divided by total loaded DNA. The designed vector contained information for a Cas9 enzyme, a CRISPR gRNA and an mCherry reporter. This in an attempt to explore the suitability of the vehicle in the delivery of gene editing systems. The release of DNA was also tested in vitro, by exposing the loaded systems to a medium with pH values in the range of 8.5 to 6. In this case, the efficiency was calculated as the total DNA released divided by the loaded DNA. The obtained nanoconjugates were characterized by UV-Vis spectrophotometer, dynamic light scattering (DLS), thermogravimetric analysis (TGA), Fourier transform infrared spectroscopy (FTIR), and scanning electron microscope equipped with energy dispersive spectroscopy (SEM+EDS). Furthermore, particle size was determined by DLS.

Figure 1.b shows the FTIR spectra of the core-shell nanoparticles, the core-shell+pDMAEMA conjugate and the free pDMAEMA. The distinctive peaks of pDMAEMA were observed on the conjugate, thereby confirming successful immobilization. Additionally, EDS analysis confirmed the presence of silver and the polymer as evidenced by the presence of silver and nitrogen, respectively (Figure 1.c). DLS results the magnetite core had a diameter around 124 nm, with an increase in particle diameter to 342 nm after coating with silver.

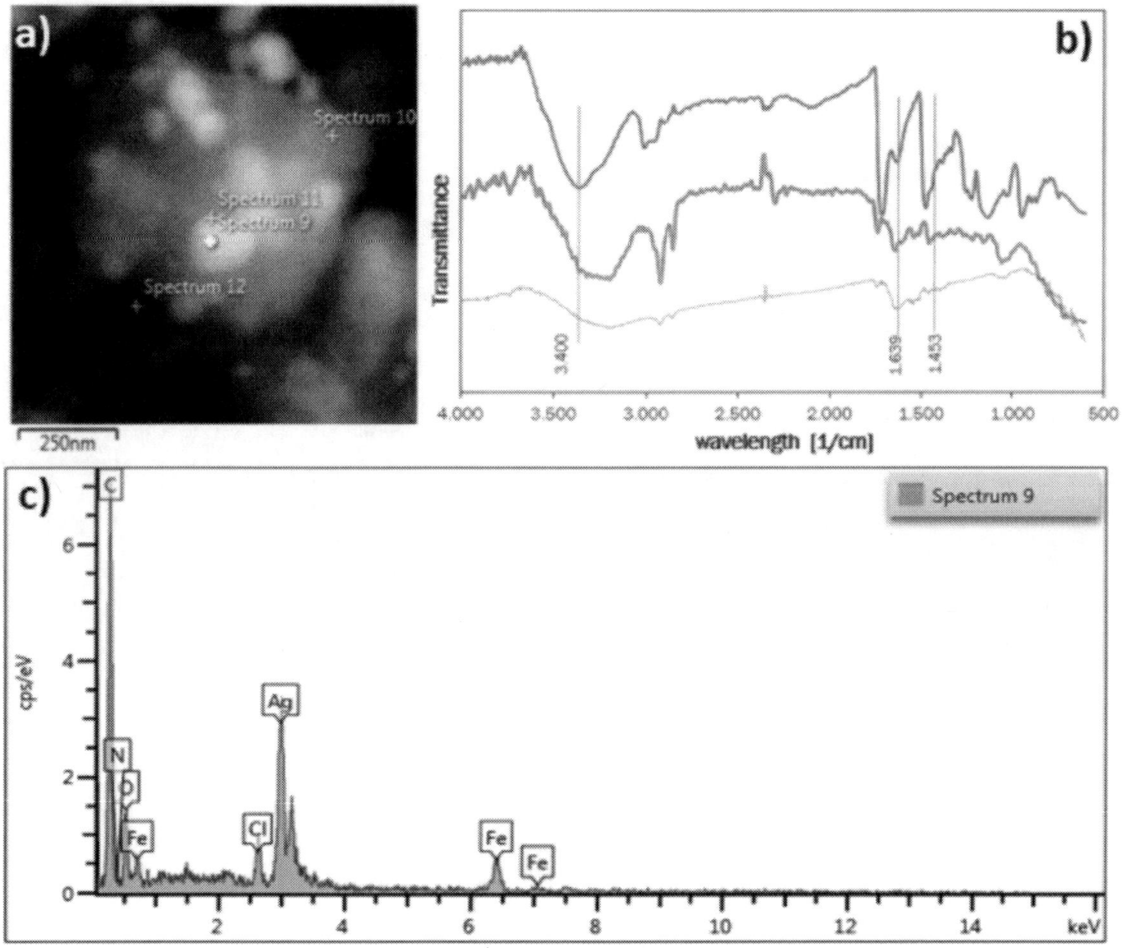

Figure 1. (a) Scanning electron microscopy (SEM) image of a core-shell nanoparticle with pDMAEMA in the surface. (b) FTIR results green is core-shell nanoparticles, red is core-shell nanoparticles with the polymer and blue is the polymer. polymer shows characteristic peaks at 1453, 1639 and 3400 [1/cm]. (c) EDS spectrum of core-shell nanoparticle + pDMAEMA shows the presence of atoms as spotted in (a) .

The study presented here provides a route for the development of gene delivery systems based on core-shell magnetic nanoparticles, and pH-responsive polymers. Also, it highlights a methodology for the synthesis, functionalization, and loading of a circular DNA vector with the size typical of those for applications in CRISPR/Cas9 editing systems. Finally, the developed system is likely to provide opportunities to overcome issues regarding low stability and endosome formation, which are commonly present when delivering with the aid of nanostructured materials.

Keywords: gene delivery, core-shell, magnetite, pDMAEMA, nanoparticles, surface modifications

References:

1. Ho, B. X., Loh, S. J. H., Chan, W. K. & Soh, B. S. In vivo genome editing as a therapeutic approach. International Journal of Molecular Sciences 19, (2018).

2. Jin, S. & Ye, K. Nanoparticle-mediated drug delivery and gene therapy. in Biotechnology Progress 23, 32–41 (2007).

3. Varadan, V. K. Nanomedicine : design and applications of magnetic nanomaterials, nanosensors, and nanosystems LK - https://univdelosandes.on.worldcat.org/oclc/916159742. TA - TT - (Wiley, 2008).

4. Cuellar, M. et al. Novel BUF2-magnetite nanobioconjugates with cell-penetrating abilities. Int. J.

Nanomedicine Volume 13, 8087–8094 (2018).

561098 Folate Receptor-Targeted Anionic Ternary Polyplexes for Gene Delivery

Nonviral gene delivery has been investigated as a promising alternative to viral vectors due to its decreased immunogenicity and safety concerns. Polymer-mediated delivery has been widely studied due to ease of polymer modification and production, potential for biocompatibility, and lack of immunogenicity, but these systems fall short of recombinant viruses in terms of gene delivery efficiency. Polymeric gene delivery vectors most commonly comprise electrostatic complexation of polycationic polymers and plasmid DNA to form polyplexes. These complexes are effective for in vitro transfection but, due to excess positive charge, often aggregate with serum proteins and interact strongly with cellular membranes leading to cytotoxicity and non-specific interactions that can interfere with cell-specific targeting. To address these problems, ternary polyplexes have recently been developed that incorporate a polyanionic component to inhibit serum agglomeration and reduce cytotoxicity. The negative charge of these polyplexes also allows the opportunity for addition of targeting ligands to promote receptor-mediated uptake, thus increasing gene delivery efficiency and specificity.

We are investigating the transfection and cellular internalization of folate-targeted ternary polymeric gene delivery vectors comprising 25-kDa branched polyethylenimine (PEI), 15-kDa poly(glutamic acid) (PGA), and pGL3 luciferase reporter plasmid. Folate was conjugated to PGA through carbodiimide crosslinking chemistry at ratios of 3, 7, and 10 folates per chain (PGA-Fol(3, 7, 10)). These conjugated polyanions were then electrostatically complexed with PEI/pGL3 binary complexes to form negatively charged ternary polyplexes. Cell lines overexpressing and under-expressing folate receptor (HeLa and HEK 293, respectively) were transfected with these ternary polyplexes and transgene expression and particle uptake were compared to non-targeted ternary polyplexes of equivalent PGA/PEI/DNA weight ratios. Transfection with PGA-Fol(7)/PEI/DNA polyplexes in HeLa cells increased uptake and transgene expression ~9-fold when compared to nontargeted polyplexes of the same weight ratio. To verify folate receptor targeting, expression of the folate receptor was knocked down through RNA interference (RNAi) and uptake was quantified via flow cytometry. The internalization mechanism was also investigated using endocytosis inhibitors and RNAi to verify specificity of folate-receptor targeting. Transfections in the presence of endocytosis inhibitors confirmed caveolin-dependent endocytosis as a primary internalization mechanism, as expected of folate-targeted internalization. Macropinocytosis was also found to be an important pathway, likely due to the efficient, nonspecific uptake of the pathway. Folate-conjugated anionic complexes improved uptake and gene expression over non-targeted particles of the same PGA/PEI/DNA weight ratio. These results suggest folate-receptor targeting as an effective system to improve the gene delivery efficiency of anionic ternary polyplexes.

570807 Nanoparticle Carrying T Cells for Targeted Drug Delivery

Bionanotechnology (22B 02)

Nanoparticle Carrying T Cells for Targeted Drug Delivery

Muhammad Raisul Abedin[a], Yongjian Liu[b], Sutapa Barua[a,1]

[a] Department of Chemical and Biochemical Engineering, Missouri University of Science and Technology,

Rolla, MO-65409

[b] Mallinckrodt Institute of Radiology, Washington University School of Medicine, Washington University in St. Louis, St. Louis, MO-63110

[1] Corresponding Author: Email: baruas@mst.edu

Abstract

Objective
The overall goal of this article is to explore a combinatorial approach in exploiting the specificity of antibody conjugated drug nanorods with the direction of polyclonal T cells for breast cancer therapy. We hypothesize that the attachment of Trastuzumab-conjugated paclitaxel drug nanorods to natural T cells will enhance their circulation time and substantially alter their anti-tumoral efficacy.

Introduction
Major limitations of the current nano drug delivery vehicles are inefficient transport of drug to the tumor site, uncontrolled nanoparticle-biomolecule interaction at the extra cellular matrix, and intracellular lysosomal or phagolysosomal degradation of the nanoparticle. Moreover, the majority of the designed nano drug vehicles accumulate in the vicinity of the off target cells or tissues resulting in off target toxicity or might be eliminated by the residing macrophages in the organs like liver and spleens. To address this issue, this project aims to take a combinatorial approach of immunotherapies, and targeted drug NPs for efficient delivery of therapeutic payloads to the breast cancer site, and expand therapeutic benefits. Here, we propose an alternative strategy that makes use of the ability of natural immune T cells to circulate and target breast cancer cells for the effective delivery of synthetic drug NPs. Specifically, it is proposed to anchor Trastuzumab conjugated paclitaxel drug nanorods to the surface of patient derived T cells as a means of avoiding the reticuloendothelial system (RES) clearance. We hypothesize that the attachment of Trastuzumab-conjugated paclitaxel drug nanorods to natural T cells will enhance their circulation time and substantially alter their anti-tumoral efficacy.

Materials and Methods

Paclitaxel nanorods (503.42±210 nm) were prepared using the solvent diffusion method. Trastuzumab (TTZ) (Genentech) was conjugated with the functionalized paclitaxel drug nanorods. The antibody drug conjugate (PTXNR-TTZ) was further conjugated with the DOTA chelator for radiolabeling with positron emitting ^{64}Cu isotope. Cytotoxic T lymphocytes were bound with PTXNR-TTZ in serum free un-supplemented RPMI 1640 medium.

Results

The SEM images confirm the cylindrical or rod-like the shape of the paclitaxel drug particles in the nano dimension of (503.42±210) × (96.87±33)nm. The nanoparticle synthesis efficiency was 65%. Monoclonal antibody trastuzumab was chemically conjugated with the paclitaxel drug nanorods with the conjugation efficiency of 53%. The drug NRs along with the antibody conjugated PTXNR-TTZ showed enhanced stability in the biological media. PTXNR-TTZ was bound on the surface of the cytotoxic T lymphocytes without inducing apoptosis to the T cells. T- cell surface bound antibody conjugated drug nanorods induced apoptosis in HER 2+ breast cancer cells resulting in significant tumor clearence.

573632 Evaluating Toxicology, Immunogenicity, Bio-Distribution, and Pharmacokinetics of DNA Origami Nanostructures *In Vivo*

Scaffolded DNA origami nanotechnology allows for the generation of pre-defined shaped nanoscale objects via molecular self-assembly as well as a robust platform for drug delivery applications. Previous studies reported that DNA origami nanostructures could be functionalized with targeting moieties and effectively loaded with anthracyclines (e.g. doxorubicin and daunorubicin) and thrombin. Furthermore, drug-loaded DNA origami nanostructures were reported to induce an enhanced anti-cancer effect relative to free drug in both solid tumor and hematologic model systems *in vitro*, while recent findings showed that thrombin-loaded nanostructures induce tumor necrosis and inhibit tumor growth *in vivo* in a targeted manner. Despite this exciting promise, the toxicology, immunogenicity, bio-distribution, and pharmacokinetics of DNA origami nanostructures *in vivo* remain ill defined. Here we evaluate two DNA origami nanostructures that vary in shape, a flat-2D triangle and 3D-rod-shaped nanostructure either alone or functionalized with polyethylene-glycol (PEG)-conjugated oligonucleotides. A repeat dosing regimen of unloaded triangle and rod-shaped structures at clinically relevant levels revealed that DNA origami nanostructures were generally non-toxic *in vivo* as shown by weight and a complete biochemical panel assessing liver function apart from mild hepatic necrosis in mice treated with PEG-coated triangle nanostructures. A modest pro-inflammatory molecular and cellular immune response was evident among mice in all treatment groups, especially among mice treated with triangle/triangle-PEG that dampened by the conclusion of the dosing regimen. Distribution findings showed PEG-functionalized and non-functionalized DNA nanostructures distribute throughout the periphery immediately after i.v. injection and within ~30 minutes after i.p. injection, while myeloid cell populations preferentially internalized DNA nanostructures *ex vivo*. PEG-functionalized DNA nanostructures demonstrated modest increased stability *in vivo* relative to unmodified DNA nanostructures via a fluorescence-based live animal *in vivo* imaging system (IVIS). Taken together, our *in vivo* findings suggest that DNA origami nanostructures are mostly non-toxic, generate a modest immune response, distribute effectively into the periphery, and, therefore represent a promising novel platform for future cancer drug delivery studies *in vivo*. *Funded in part by start-up funds provided to Dr. Castro by the Department of Mechanical and Aerospace Engineering, The National Institutes of Health-National Cancer Institute-P50-CA140158 and R35-CA197734, The D. Warren Brown Foundation, Four Winds Foundation, and The Harry T. Mangurian Jr. Foundation. C.R.L is a recipient of a National Institutes of Health T32 Award in Oncology Training Fellowship at The Ohio State University Comprehensive Cancer Center, T32-CA009338.*

566459 Rational Selection of Intestinal Permeation Enhancers for Oral Macromolecular Drug Delivery

Oral delivery of proteins remains a pressing, unmet need in the field of drug delivery, primarily due to poor transport across the intestinal epithelium. A common approach to achieve transepithelial transport is the use of materials called permeation enhancers that manipulate the epithelial cell monolayer. Unfortunately, the clinical translation of permeation enhancers has been limited. Although many reported enhancers increase drug permeability across commonly-used Caco-2 cell monolayers, their efficacy rarely extends to studies in mice or rats. Poor translation from in vitro to in vivo systems can be partially attributed to inadequate understanding of the physical limits of intestinal permeation enhancement in cell culture and in animals. This study systematically determined how three mechanistically distinct permeation enhancers affect the in vitro and in vivo permeability of macromolecules with varied size, surface charge, and conformation. This newfound knowledge will enable rational selection of permeation enhancers when developing delivery systems for therapeutic proteins.

Five proteins (insulin, 5.7 kDa; cytochrome C, 12 kDa; ovalbumin, 44 kDa; bovine serum albumin, 66 kDa; and mouse IgG, 150 kDa) and five dextrans with corresponding molecular weights were fluorescently labeled and characterized for hydrodynamic diameter, molecular weight, and surface charge. Permeability was measured in vitro using Caco-2 monolayers and in mice by intestinal injection. Solutions of one marker mixed with one of three enhancers (1-phenylpiperazine, silica nanoparticles, or sodium deoxycholate) were administered to cells or mice, and fluorescence measurements quantified permeation from media or blood samples.

The three permeation enhancers operated via different mechanisms in Caco-2 monolayers, and the effect of macromolecular size on permeation varied significantly with enhancer type. In Caco-2 monolayers, silica nanoparticles were efficacious for markers less than 40 kDa, while low concentrations of 1-phenylpiperazine increased permeability of all markers. Translation to mice was challenging in that effective in vitro enhancer concentrations were not reflective of required concentrations in vivo. However, with optimization of enhancer doses, significant permeability increases were observed in mice. For 4 kDa dextran, silica nanoparticles more than doubled the permeability and 1-phenylpiperazine achieved a remarkable 17-fold increase in permeability.

This study improves our understanding of how permeation enhancers aid intestinal absorption of macromolecular drugs by thoroughly examining the performance of three mechanistically distinct permeation enhancers. Ultimately, compatibility between enhancer mechanism and the properties of the macromolecules dictated permeation enhancer performance in mice.

575928 Chiral Supraparticles for Controllable Nanomedicine

Chirality is ubiquitous in nature that is hard-wired into every living biological system. The well-known examples are *L*- type amino acids and *D*- type sugars such as DNA and RNA. Since the chirality of building block decides the handedness of assembled structures, proteins such as enzymes are also chiral. A volatile molecule limonene may smell like pine or lemon depends on the handedness of the molecule that reaches to nose. For the same reason, a substrate may or may not be an effective medicine, and makes foods taste bitter or sweet because all receptors in our nose, cells, and tongue in general are chiral and react differently depends on the enantiomer they encounter.

Despite of the critical role, the nexus of chiral bioengineering has not been explored. One of them is lipids in cellular membrane that are all right-handed. Lipids are the major component that decides biophysical properties of cell membrane. Hence lipid composition is the first metabolic change with most of human diseases including metabolic, immune and cancers. It has been well demonstrated that tumor/cancer cells possess increased amount of cholesterol in the cell membrane. When tumor expresses excess amount of cholesterols, it adapts drug resistance because it promotes highly ordered and rigid cell membrane. The design of nanomedicine to overcome this barrier is significantly important for effective drug delivery.

Chiral nanostructures are a sharp tool because cells possess one type of lipid enantiomer in the membrane. It has been well understood that interactions between *L*- to *L*- and D- to *D*- are thermodynamically stronger than *L*- to *D*-. Here we designed chiral supraparticles (SPs) that react distinctively to cell membrane depends on their handedness. SPs coordinated with *D*- chirality showed at least four times stronger cell membrane penetrations and anti-cancer properties. We carried out QCM-D and ITC measurements to understand the mechanism, which confirmed that *D*- SPs has more effective adhesion on lipid layers than *L*-SPs. When it comes to *in vivo* environments that contains a large, heterogeneous population of proteins, *D*-SPs showed superior stability and longer biological half-lives due to the incompatible chirality with endogenous proteins including proteases. This study shows that incorporating *D*-chirality into nanosystems enhances cellular uptake and *in vivo* stability in blood providing support for the importance of chirality in bioengineering. As a result, chiral engineering may have the potential to provide a new level of control for DDS, tumor detection markers, biosensors, and many other biomaterial devices.

563147 Quantifying Nanoparticle Delivery to Solid Tumors at the Cellular Level

In this presentation, we will explore interactions of engineered nanoparticles within the tumor microenvironment. Using quantitative analytical techniques, such as flow cytometry and inductively-coupled plasma mass spectrometry, we determined the delivery of systemically administered nanoparticles to individual tumor cell populations. Our results revealed that non-cancerous cells, such as tumor associated macrophages, represent biological barriers to efficient nanoparticle delivery to cancer cells in solid tumors.

577026 Aloe Vera-Mediated Green Synthesis of Tellurium Nanostructures with Both Antimicrobial and Anticancer Activity

Antimicrobial resistance to antibiotics is a huge concern that presents significant challenges to modern healthcare. Thus, an alternative solution that does not require the use of antibiotics is needed. One of the most promising methodologies comes from the implementation of nanotechnology. Unlike antibiotics, bacteria are unable to develop resistance to nanoparticles. Many methods for synthesis of various types of nanostructures have been reported. Nevertheless, they are often accompanied by significant drawbacks such as cost, harsh processing conditions, and production of toxic by-products that are a concern for both the environment and society. In this work, an environmentally-friendly and cost-effective approach for synthesis of tellurium nanorods in aqueous media have been developed using aloe vera extract as a unique reducing agent. Nanorods were characterized using Transmission Electron Microscopy and Energy-Dispersive X-Ray Spectroscopy to determine size, morphology and composition. Resulting nanoparticles with a length 100±19 nm and width of 5±2 were produced. Nanoparticles were also characterized and tested for their ability to inhibit bacterial growth. A decay in bacterial growth after 24 hours was achieved for both *Staphylococcus aureus* and *Escherichia coli* at tellurium nanoparticle concentrations from 5 to 75 μg/mL. Biocompatibility assays were also performed, and the nanoparticles showed no cytotoxic effects for human dermal fibroblast cells after 48 hours. Anticancer properties were tested as well, and results indicated that the nanoparticles (in concentrations ranging from 5 to 75 μg/mL) were able to delay cancer cell proliferation. Aloe vera extracts afforded synthesis of tellurium nanoparticles in a quick, cost-effective and completely green approach that showed both antimicrobial and anticancer effects with low cytotoxicity for healthy cells.

567665 Cancer Microtissues for Rationalized Nanoparticle Based Radiotherapy

Radiotherapy is an integral and highly effective part of cancer therapy applicable in over 50% of the patients affected by cancer. Due to the low specificity of irradiation, the maximal dose is greatly limited in order to not cause damage to healthy tissue. The limited dose oftentimes leads to the survival of a small portion of radio-resistant cells that then grow back causing the cancer to reoccur. Approaches based on high z (atomic number) inorganic nanoparticles have been proposed to locally increase the photoelectric absorption cross-section relative to healthy tissue. However, the complex interplay between the radio-enhancers and the tumor tissue has led to radio-sensitizer drug treatment plans based mostly on empirical decisions. Here, we report on a systematic approach to rationalize radio-enhancer development and combination therapy plans based on quantitative endpoints. We present an advanced in vitro model depicting key features of tumors and report on the nanoparticle uptake and radio-enhancement effects compared to 2D in vitro culture.

L.H.R. Gerken, K. Keevend, Y. Zhang, F.H. L. Starsich, C. Eberhardt, G. Panzarasa, M.T. Matter, A. Wichser, A.s Boss, A. Neels, I.K. Herrmann, Lanthanide-doped Hafnia Nanoparticles for Multimodal Theranostics: Tailoring the Physicochemical Properties and Interactions with Biological Entities, ACS Applied Materials and Interfaces, 2019.

573270 Projection-Specific Modulation of Neural Activity with a Non-Genetic Method

Recent developments in optogenetics and viral technology have demonstrated that modulating the neural activity of specific projections in the brain is capable of rescuing behavioral deficits, including those relevant to depression, autism and anxiety. However, this approach requires genetic modification of neurons via viral transduction, which introduces a significant barrier for clinical applications due to safety concerns. Also, the current projection-targeted neural modulation methods all use visible light, limiting both the light penetration depth and the number of independent channels neuroscientists can use for recording and modulating neural activity. Here, we are developing a non-genetic, nanoparticle-based approach to achieve projection-specific modulation of neural activity in freely behaving mice. First, we observed that gold nanorods were rapidly endocytosed into cultured neurons after just a few minutes of incubation. Transmission electron microscopy images illustrate that gold nanorods are located both inside and outside the endosome. Second, we showed that the endocytosed gold nanorods were transported retrogradely/anterogradely along neuron axons, with a room temperature average speed of 0.2 μm/s (1.7 cm/day). Third, we demonstrated the effectiveness of photothermal inhibition of neural activity with axon-transported gold nanorods, illustrating the concept of projection-specific modulation of neural activity *in vitro*. The mechanism of photothermal neural modulation is likely due to the thermal response of potassium channels, as shown from our *in vitro* electrophysiology studies. In addition, we demonstrated the excellent biocompatibility of gold nanorods with neurons after at least one-week incubation and photothermal neural modulation. Finally, we observed the neural uptake and axonal transport of gold nanorods *in vivo,* as well as neural modulation in mice with this technology. Overall, our nanoparticle-based methodology demonstrates a promising approach towards non-genetic, projection-specific modulation of neural activity.

575445 Understanding and Controlling the Plant Antifungal Activity of New Classes of Environmentally Benign Nanoparticles

Our group has introduced a class of environmentally benign nanoparticles (EbNPs) for biotechnology (Nature Nanotechn. 10:817, 2015). These nanoparticles made of lignin can biodegrade after use, with low potential for nanotoxic impact. We use a semi-continuous synthesis system to produce large amounts of EbNPs by a non-LaMerian mechanism, and we investigate these nanoparticles as delivery vehicles of actives to biological surfaces. These EbNPs have already been proven to be efficient antibacterial agents when loaded with silver ions and coated with polyelectrolyte. We are now using the biopolymer chitosan as a nanocoating, allowing us to control the surface charge of the EbNPs. The charge due to chitosan enables the EbNPs to adhere to plants, which usually have an intrinsic negative charge. We evaluate the ability of uncoated and coated EbNPs to prevent damage from the fungus *Botrytis* to roses compared to the industry standard chemical fludioxonil. While chitosan coated nanoparticles have improved adherence to petals, the uncoated EbNPs prevent fungal growth more efficiently. The lignin nanoparticles appear to protect the petals by disrupting the interaction between them and spores, and by adhering to spores, impeding their development and causing aggregation. Using brightfield, fluorescence, and scanning electron microscopy, we study the fundamental interactions between spores, EbNPs, and rose petals to understand the origins of the antifungal activity of the lignin nanoparticles. We observe the effects of EbNP treatment before and after the inoculation of rose petals with spores to understand how the timing and physical characteristics of lignin nanoparticles and spores impact growth and development of germ tubes and hyphae. We will report our conclusions about the correlation of the colloidal interactions, "coffee ring" formation, and infectivity to rose petals. By understanding these interactions, one can formulate novel, efficient, and environmentally benign antifungal products.

573133 Effect of Laser Irradiation on Drug Release and Nano-Structural Reversibility of Light-Activatable Methotrexate-Encapsulated Liposomes

Drug delivery systems for temporal and spatial drug release using nanocarriers, or so-called on-demand drug delivery systems, are a new therapeutic approach in the medicinal field. Not only can drug release be tightly controlled and negative side effects associated with incorrect dosage minimized, but also the therapeutic efficacy can be prolonged by conserving drug when not needed.[1-2] A photoactive-triggering mechanism of site-specific drug delivery has gained interest within this field of research.[3-4] Although the technique shows potential for repeatable on-demand drug delivery, the lack of understanding of nanostructural changes of the drug-carriers with various parameters of light activation often limit repeated application.

Here we report on novel plasmonic material-coated liposomes (of mixed lipids) for the efficient delivery of methotrexate (MTX), a common drug for cancer and autoimmune diseases. Gold nanorods are used as the plasmonic material on the liposomes, which generate heat during near-infrared laser irradiation. The repetitive drug release from these MTX-encapsulated liposomes was measured using UV-Vis spectroscopy. The reversibility of the liposome structure after repetitive laser irradiation was determined using a single particle electrochemical measurement which measures the frequency of liposome collision events via changes in current that occur when a liposome collides with, and blocks, an electrode surface (2 μm diameter Pt disk).

We find that drug release occurs with each repetitive laser irradiation, whereas the liposome collision frequency (and thus concentration of liposomes) does not change. The collision frequencies of the irradiated liposomes were 0.056 Hz, 0.060 Hz and 0.060 Hz for 0 s, 30 s and 120 s irradiation time, respectively (1.9 W average power) indicating that the liposomes remain intact after irradiation. However, when MTX-encapsulated nanodroplets that underwent phase-transition from liquid to gas bubbles[5] were irradiated, the collision frequencies were 0.13 Hz, 0.107 Hz and 0.023 Hz, for 0 s, 30 s and 120 s, respectively, demonstrating 18% and 82% decrease in collision frequencies for 30 s and 120 s irradiation, respectively. The results indicate the number density after irradiation was maintained for the liposomes but not for the nanodroplets, suggesting reversibility of the liposome structure.

When larger liposomes (mean 3 μm in diameter) were irradiated, the mean current drop per collision increased from 0.2 nA to 0.8 nA, compared to the regular liposomes tested above (230 nm in average diameter). However, the collision frequencies were 0.093 Hz, 0.09 Hz, and 0.103 Hz for 0 s, 30 s and 120 s, respectively, again indicative of the structure reversibility.

The results suggest that the gold nanorod-coated liposomes can be a potential drug carrier for repetitive on-demand drug delivery. In addition, we expect to measure drug dosage released from an individual carrier, leading to precise control of dosage multiple times for long term drug delivery.

References

1. Manzoor, A. A.; Lindner, L. H.; Landon, C. D.; Park, J.-Y.; Simnick, A. J.; Dreher, M. R.; Das, S.; Hanna, G.; Park, W.; Chilkoti, A., Overcoming limitations in nanoparticle drug delivery: triggered, intravascular release to improve drug penetration into tumors. *Cancer research* **2012,** *72* (21), 5566-5575.
2. Poznansky, M. J.; Juliano, R. L., Biological approaches to the controlled delivery of drugs: a critical review. *Pharmacological reviews* **1984,** *36* (4), 277-336.
3. Dougherty, T. J.; Gomer, C. J.; Henderson, B. W.; Jori, G.; Kessel, D.; Korbelik, M.; Moan, J.; Peng, Q., Photodynamic therapy. *JNCI: Journal of the national cancer institute* **1998,** *90* (12), 889-905.
4. Jin, Q.; Mitschang, F.; Agarwal, S., Biocompatible drug delivery system for photo-triggered

controlled release of 5-fluorouracil. *Biomacromolecules* **2011,** *12* (10), 3684-3691.
5. Zhang Z., Taylor M., Collins C., Haworth S., Shi Z., Yuan Z., He X., Cao Z., Park Y. C., Light-activatable theranostic agents for image-monitored controlled drug delivery. *ACS Appl. Mater. Interfaces*, **2018**, *10* (2), 1534–1543.

573134 Near Infrared, One-Photon Triggered Liposome Cage for Calcium, ATP or Both

We describe a one-photon, near infrared triggered carrier for calcium, adenosine triphosphate, and potentially any small water-soluble bioactive molecule using unilamellar liposomes tethered to plasmon-resonant hollow gold nanoshells (HGN). Picosecond or shorter pulses of NIR light, available in conventional two-photon microscopes, at a fluence dependent on the plasmon resonance, size, and liposome composition, are adsorbed by the HGN, which rapidly heat, then dissipate that heat by forming vapor nanobubbles that ruptures the liposome carrier, releasing the calcium, ATP or other liposome contents within milliseconds. The spatial resolution is set by the size of the liposome carrier, as is the amount of encapsulated molecule released. This carrier system is universal and can release any small molecule at the same rate and in the same fashion, unlike current chelating cages for ions or photolabile protecting groups for biomolecules that require different chemical synthesis to achieve. By using liposomes of different composition and HGN with different nanobubble threshold fluences, two different caged molecules can be released simultaneously, or one before the other, or in a desired ratio. The vesicle and the HGN are biocompatible and provide complete and stable separation of the bioactive compound from its surroundings until triggered by NIR light. We will present our initial findings of release in various cell lines.

563158 Cholesterol Content Mediates Tumor Cell Uptake of Targeted Liposomes

Effective design of nanocarriers is incumbent upon thorough understanding of material properties and their interactions at biological interfaces. Within this context, the role of nanoparticle mechanical properties on cellular uptake is poorly understood. By altering membrane fluidity via cholesterol composition, we show that liposome cholesterol content controls particle uptake kinetics of CD44-overexpressing high grade serous ovarian cancer cells. Hyaluronic acid-coated liposomes were synthesized with membrane cholesterol mole compositions ranging from 0% to 40%. Liposome bilayers were fluorescently tagged with NBD and the core was loaded with calcein blue to differentiate uptake of lipophilic and hydrophilic cargo respectively. Uptake kinetics of the NBD-calcein blue nanoparticles was assessed via fluorescence-activated cell sorting (FACS). At early time points (4hr), particle accumulation of layered liposomes containing no cholesterol was found to be significantly higher than layered liposomes containing 8%, 30% and 40% cholesterol ($P < 0.001$). These differences in particle uptake kinetics were confirmed via fluorescence confocal microscopy. Uptake mechanisms were probed using a panel of inhibitors. ATP-depletion and inhibition of receptor-mediated endocytosis prevented uptake of layered liposomes comprised of 40 mol% cholesterol, but did not prevent uptake of layered liposomes containing 0 mol% cholesterol. This work demonstrates that cholesterol composition can be used to control the uptake kinetics and uptake mechanism of targeted liposomes.

575941 Serum Effects on Nanoparticle-Mediated Photoporation for Enhanced Intracellular Drug Delivery

Intracellular delivery of therapeutic and diagnostic molecules is restricted by plasma membrane. Often, endocytic route is used to transport molecules inside cells, which can render these molecules inactive due to pH changes. Nanoparticle-mediated photoporation offers a physical route to create transient pores allowing uptake of foreign substances by cells. Through near-infrared laser irradiation, nanoparticles absorb and dissipate energy to the surroundings, thereby vaporizing water to create steam bubbles. The subsequent thermal and acoustic outputs are believed to be responsible for membrane poration. Typical *in vitro* experiments are conducted with DU145 prostate cancer cells, carbon black nanoparticles and fluorescent uptake markers in cell suspension media, kept inside a cylindrical glass cuvette irradiated with 1064 nm wavelength laser beam. Post irradiation, cells are stained with PI (propidium iodide) for viability loss measurement and subsequently washed to remove nanoparticles and excess staining molecules. Cells are then analyzed for uptake and viability loss using flow cytometer.

This process has been optimized for >90% of cells exhibiting intracellular uptake of low-molecular weight (less than 1 kDa) marker compounds without significant cell viability loss. Understanding how changes in cellular micro-environment affect delivery efficiency is important for practical translation of this platform technology. Therefore, this study is focused on understanding the role of serum during photoporation, and how that understanding can be used to increase intracellular delivery of macromolecules.

Experimental results reveal 75% less loss of cell viability during laser irradiation at high fluence when cells are suspended in media containing 10 (v/v)% serum. Similar effects are observed in media containing albumin (major protein component present in serum) as well as denatured serum proteins. Further experiments show that a few polymer additives also help preserve cell viability during photoporation and thus viability protection by serum appears to be attributed to physical interactions between serum components and cell membrane/suspension media and not to biological activity, for example, associated with serum proteins. The ability of cells to stay alive at high fluence may enable macromolecular delivery (greater than 10 kDa) which have applications in protein and gene therapy. Thus, future experiments will explore these avenues in addition to investigating the mechanism behind cell viability protection in presence of serum and polymer additives.

Reference:

1. Sengupta et al. Efficient Intracellular Delivery of Molecules with High Cell Viability Using Nanosecond-Pulsed Laser- Activated Carbon Nanoparticles. *ACS Nano* 2889–2899 (2014)

567404 Autophagy Targeting Nano-Prodrugs for Synergistic Therapy of Metastatic Cancer

Cancer metastasis accounts for over 90% of the carcinoma-related mortality. During metastatic dissemination, metastatic tumor cells may suffer from anoikis, a form of programmed cell death happened in extracellular matrix (ECM)-detached cells. However, autophagy induction can prevent tumor cells from anoikis by multiple mechanisms. In addition, disassembly of focal adhesions by autophagic degradation contributes to the metastatic progression. Here, we constructed redox-responsive dual-polymeric-prodrug nanoparticles delivering SN38 (active metabolite of irinotecan) and hydroxychloroquine (HCQ, an FDA approved autophagy inhibitor) with optimal synergetic ratio. The combo nanoparticles (Combo NPs) containing PEG-PSN38 and PEG-PHCQ at 1:5 molar ratio displayed synchronized release in the reductive tumor environment, and remarkably improved the pharmacokinetic behaviors and tumor accumulations of both SN38 and HCQ. Importantly, the Combo NPs displayed efficient inhibition of tumor growth and metastasis using 4T1-bearing orthotopic and lung metastatic breast cancer models. Mechanism studies indicated that autophagy inhibition by the Combo NPs could attenuate the migration and invasion of 4T1 cells through impairing anoikis resistance and degradation of paxillin (a key component of focal adhesions). Therefore, the reported polymeric nano-prodrugs rationally combined autophagy inhibition and chemotherapy, representing a potential therapeutic approach for blocking metastatic dissemination of breast cancer.

576678 Engineering pH Responsive, Self-Assembled Polymer Nanoparticles for the Oral Delivery of Protein Therapeutics

Engineering pH Responsive, Self-assembled Polymer Nanoparticles for the Oral Delivery of Protein Therapeutics

Matthew K. Miller, Michael G. Elder, Lexi C. Lakatta, and Nicholas A. Peppas

Introduction: While protein therapeutics are promising treatment options in many realms of disease, they are limited by their fragility. These therapeutics must be delivered parenterally, which can lead to reduced patient compliance. An oral delivery mechanism would be preferable; however, the conditions of the human gastrointestinal tract are not amenable to naked protein delivery. Thus, this work focuses on the development of a self-assembled, pH responsive nanoparticle system capable of delivering antibodies orally comprised of block co-polymers of methacrylic acid (MAA), methyl methacrylate (MMA), and PEG.

Methods: A reversible deactivation radical polymerization scheme was used to synthesize poly(trimethylsilyl methacrylate-co-methyl methacrylate), which was then conjugated to linear maleimide functionalized PEG10000 via carbodiimide coupling chemistry. The trimethylsilyl groups were removed in mild conditions to yield the final pH responsive di-block copolymer. Polymer compositions and molecular weight were quantified by NMR and FTIR. Nanoparticles were formed using a nanoprecipitation technique, and size and pH mediated stability were analyzed via dynamic light scattering and nanoparticle tracking analysis. Drug loading and release from the nanoparticles were analyzed *in vitro* using the model protein IgG. IgG concentrations were measured by ELISA.

Results: The co-polymers exhibit a pH responsiveness mediated by the pKa of MAA, which can be adjusted by co-polymerization with MMA. At a ratio of 1:1 MMA:MAA, the polymer exhibits a pKa of approximately 6.0. This pKa allows the nanoparticle to remain assembled throughout the stomach and upper small intestine where the pH ranges from approximately pH 1-6. After passing through the intestinal epithelium, the particle disassembles in the physiological pH conditions found in the lamina propria, releasing its payload. *In vitro* drug release studies indicate the particles release roughly 10-15% of their protein payload in pH 2.0 and pH 6.0 buffers, with near 100% release in pH 7.4 conditions. The formed nanoparticles have a z-average diameter from 150-200nm with drug loading efficiencies greater than 70% for polymer loaded in a 10 wt% IgG solution. Future work will incorporate targeting moieties on the nanoparticle surface via maleimide-thiol chemistry to facilitate transepithelial transport.

Conclusions: The block co-polymers synthesized in this work appear to be viable candidates for the oral delivery of protein therapeutics. The self-assembled particles protect the drug in the stomach and upper small intestine, while the higher pH of the lamina propria leads to disassembly of the particle and release of the therapeutics. Particle disassembly also facilitates more efficient release of the payload compared to other delivery systems.

Acknowledgements: This work was supported in part by the National Institutes of Health (R01-EB-000246, R01-EB-022025) and the Cockrell Family Regents Chair. M.K.M. was also supported in part by a National Science Foundation Graduate Research Fellowship (DGE-1610403), and the Jodie Isenhower Endowed Graduate Fellowship in Engineering.

577766 Systemic Dendrimer-Mediated Delivery of Triptolide Specifically to Tumor-Associated Macrophages Improves Efficacy in Glioblastoma

Introduction: Glioblastoma is among the most severe types of cancers due to challenges with delivering therapies to the brain in addition to traditional solid tumor barriers. The blood brain barrier acts as a natural defense mechanism preventing foreign substances, including medicines, from reaching the brain. In addition, ill-developed vasculature and high interstitial pressure within tumors hampers effective drug penetration. This two-pronged barrier requires the administration of high doses of systemic therapies to achieve clinically relevant quantities in the brain tumor, resulting in significant off-target toxicities. Triptolide is a promising anti-cancer therapeutic that acts both as an immune suppressant in tumor-associated macrophages (TAMs) and as an anti-proliferative in cancer cells. However, like many potent anti-cancer drugs, its off-target activity causes significant systemic toxicities. Therefore, a platform that carries triptolide to the brain and into the brain tumor for triggered release while remaining inactive in the body provides an optimal clinical strategy.

Methods: Our group has shown that hydroxyl-terminated polyamidoamine (PAMAM) dendrimers selectively target activated microglia and macrophages in several neurodegenerative disorders, including TAMs in brain cancers. Here, we present novel PAMAM dendrimers surface modified with triptolide via ester linkages for intracellular and intratumoral triggered release (D-Trip). *In vitro* studies were performed in GL261 murine microglia to assess anti-proliferative activity and in BV2 murine microglia stimulated with IL-4 to explore immune-suppressant activity. *In vivo* experiments were performed in the GL261 mouse model of glioblastoma. Tumor-bearing mice were systemically administered 0.5 mg/kg triptolide or D-Trip daily starting day 5 post-inoculation and monitored regularly for markers of disease progression.

Results: D-Trip released triptolide in a sustained manner over two weeks under intracellular conditions while exhibiting minimal release in extracellular conditions. *In vitro*, D-Trip showed improved immune suppression in TAMs-like activated microglia over triptolide while both performed similarly in inducing tumor cell death. With systemic administration in mice, D-Trip treatment significantly reduced tumor burden ($p = 0.04$ vs. triptolide, $p = 0.02$ vs. control), leading to improved behavioral markers of disease ($p = 0.0012$ vs. triptolide in kyphosis score). D-Trip treatment also ameliorated the significant toxicity at the injection site observed in triptolide treated animals attributed to the targeting and triggered release capabilities of the dendrimer platform.

Conclusions: Here, we present a novel dendrimer-mediated triptolide targeted delivery platform that provides multiple anti-cancer functionalities for improved therapeutic efficacy. Dendrimers specifically target TAMs and release triptolide intracellularly to suppress their pro-tumor polarization. These TAMs then act as depots to release dendrimers into the surrounding tumor microenvironment to induce tumor cell death. These studies demonstrate the significant potential of PAMAM dendrimers for overcoming challenges to brain tumor drug delivery.

576943 Solid-Phase Synthesis of Megamolecules: A New Class of Nanoscale Therapeutics

Proteins are versatile building blocks in the field of synthetic biology. Novel proteins can be engineered to perform unique functions and chemistry for both biological and industrial applications. However, synthesis of non-natural molecules greater than 100 kDa presents a large metabolic burden on bacterial hosts, leading to issues in aggregation, low yield, and purity. Most approaches for the in vitro construction of solution-phase multi-protein assemblies require high concentrations of protein building blocks, costly purification steps, and result in heterogeneous products. Taking inspiration from approaches such as solid-phase peptide synthesis, we present a synthetic strategy to produce homogeneous protein-based nanoscale structures, referred to herein as megamolecules, by immobilization of protein building blocks on SnapTag-functionalized resin beads. Building from the SnapTag capture chemistry, protein building blocks are constructed by step-wise assembly into large molecules using multifunctionalized bi-, tri-, and tetra-functional crosslinker molecules. Expanding upon the library of traditional capture chemistries, the crosslinks capitalize upon the specificity and irreversibility of the active site-ligand complex. Through the construction of a toolbox of oligomers and the paired inhibitor molecules, the creation of a library of megamolecules is possible. Upon proteolytic cleavage from the resin by an engineered TEV recognition site between the bead and the megamolecule assembly, this method produces a monodisperse product that requires no formal purification process - suggesting ease and feasibility in scale up for industrial applications. Through rational design, we generated cost-effective biomaterial therapeutics that exhibit non-linear connectivity in high yield with masses up to 500 kDa. Using this synthetic-strategy, we constructed a panel of antibody-enzyme conjugates for prodrug therapy in a breast cancer model. Leveraging the ability to control the stoichiometry of both the enzyme and antibody domains assembled on the nanoscale structure, we assembled biorthogonal megamolecules to target overexpressed HER2+ breast cancer cells in 3D cell culture. Once the megamolecule is bound to the cancer cell membrane, the turnover of the non-cytotoxic prodrug into a cytotoxic drug is localized to the cell, minimizing off target toxicity to healthy cells. We believe that the construction of antibody-enzyme conjugate megamolecules will enhance the established techniques of nanotherapeutic delivery of prodrug cargo through improvements in spatial ligand binding, which is dependent on specificity, uptake, binding, and recognition of the nanostructure. Based on the function-structure relationship of multi-protein complexes, we expect that tuning the topology of the nanostructure will enable the generation of a new class of cell-specific drug delivery therapeutics.

561684 Developing Nanofiber-Based, Safe and Functional Cell Delivery Devices for Type 1 Diabetes

Type 1 diabetes (T1D) is an autoimmune disease where the patients' insulin-producing pancreatic islet cells are mistakenly destroyed by their own immune system. Delivery of immunoprotected donor islets or more ideally stem cell-derived insulin producing cells has been considered as a promising, superior alternative to the standard, imperfect and tedious insulin therapy. However, it remains challenging to develop a cell delivery device that is long-term functional and has none or minimal safety concerns especially when stem cell-derived cells are used. In this talk, I will discuss our recent efforts to develop different types of cell delivery devices with a particular focus on nanofiber-based ones that have the potential to overcome both the functionality and safety challenges.

576939 Green Synthesis of a Synergetic Structure of Tellurium Nanowires and Metallic Nanoparticles for Biomedical Applications

Statement of Purpose: Health care system is facing significant concerns nowadays such as antimicrobial resistance and cancer. New approaches should be considered, and nanotechnology has been found as a powerful solution to them. Current synthetic methodologies for production of nanoparticles, based on physicochemical standards are known to be easy-to-get straightforward. Nevertheless, there is a cost associated with the limitations that should be overcome from these approaches, such as the production of toxic by-products or the lack of biocompatibility of the products. Therefore, new methods are needed, and green chemistry offers itself as a suitable and novel answer, achieving a safe and environmentally-friendly design, manufacture and use of chemical products. In this research, tellurium nanowires were synthesized using a green synthesis methodology (TeNWs). They were characterized in terms of structure and composition and tested for anticancer, antibacterial and cytotoxicity properties. Then, TeNWs were used for the synthesis of metallic nanoparticles in an easy and straightforward method with no need of reducing agent that was completed within 1 minute of reaction. Nanoparticles were characterized, and the synthetic process was compared with the ones described in literature, with the aim to compare the methods in terms of chemical needed, reaction conditions and economic implications.

Methods: Tellurium nanowires were prepared using a hydrothermal reaction. The environmentally-friendly approach led to the use of telluric acid and starch as a unique reducing agent. Once purified, tellurium nanowires were used as a template for the growth of metallic nanoparticles (such as platinum and palladium) in a quick method with no need of additional reducing agent at room temperature. The structure containing both metallic nanoparticles and nanowires was known as synergy. Besides, biocompatibility and anticancer tests of both structures – the synergy and the nanowires - with human tissue were accomplished, growing human dermal fibroblast (HDF) cells and melanoma cells in media in the presence of both nanosystems. After an incubation time of 5 days, the cell growth was analyzed using MTS assay. Furthermore, antibacterial properties were tested against *Escherichia Coli* and *Staphyloccocus Aureus.*

Results: It was demonstrated that green synthesized tellurium nanowires can be used as a template for the growth of metallic nanoparticles in a quick reaction that takes places in 1 minute, at room temperature with no need of stirring and reducing agent. TEM images of both tellurium nanowires and metallic nanoparticles were taken (Fig.1), showing nanoparticles with a constant distribution size attached to the tellurium nanowires, which were quickly released from the structure. The chemistry of the samples was confirmed using EDX analysis, showing the distinct peaks of tellurium and the ones corresponding to each one of the metals. In vitro cytotoxicity assays were performed with human dermal fibroblasts (HDF) cells. The experiments showed that the use of green nanostructures, enhanced the growth of the cells in comparison with the control. Furthermore, anticancer and antimicrobial studies showed an improved performance of the synergetic structure compared to the bare nanowires structures, causing a higher depletion of cell viability.

Conclusions: Current and main methods to synthesize nanostructures, both nanowires and nanoparticles, use approaches which employ chemical methods. The weakness of these procedures (extreme reaction conditions, production of toxic byproducts...) calls for a necessity of alternative approaches. Green chemistry can be used to overcome these drawbacks. Here, green-synthesized tellurium nanowires were compared with chemically-synthesized structures to show that the first ones have an enhanced biocompatibility and anticancer properties over the ones synthesized using traditional methods. Besides, it was demonstrated that tellurium nanowires synthesized using starch can be used for the controllable and quick growth of metallic nanoparticles.

577556 Graphene-Silk Fibroin and Noble Metal Nanocomposites for Biosensing Applications

Graphene-Silk Fibroin and Noble Metal Nanocomposites for Biosensing Applications

R. Kenneth Sims, Jenny Wang, Kenneth C. Brinson, Thomas A. Alvermann, Alexander N. Mitropoulos, Kamil Woronowicz, Pamela L. Sheehan, Preston C. Haney, Harry L. Moore, F. John Burpo, Enoch A. Nagelli*

Department of Chemistry & Life Science, Chemical Engineering Program

Department of Civil and Mechanical Engineering

United States Military Academy, West Point, NY 10996

[1]Explosive Ordnance Disposal, Demilitarization & Experimental Directorate, Army Futures

Command, RDECOM-ARDEC, Picatinny Arsenal, NJ

*Corresponding PI: Dr. Enoch Nagelli, **Email:** enoch.nagelli@westpoint.edu

The superior properties of graphene make it the ideal platform nanomaterial for metal nanoparticle deposition within biopolymer nanocomposites for biosensor applications. The integration of graphene and noble metal nanoparticles with biopolymers enhances the overall electrical, thermal, chemical, and mechanical properties to develop the next generation of nanoscale biosensors. Current methods to develop conductive biomaterials involve the need for external stimuli sources and the use of harsh reducing agents.[1-5] A more simple and scalable method is the electroless deposition of noble metal nanoparticles through a spontaneous galvanic displacement reaction to produce graphene-biopolymer-noble metal nanoparticle hybrid nanocomposites. The galvanic displacement process involves the formation of noble metal nanoparticles onto the silk fibroin structure through electron transfer from the differences in the reduction potential involving noble metal ions and the supporting metal substrate.[6-7] This novel and spontaneous process involves noble metal nanoparticle deposition through the immersion of copper supported silk-fibroin thin films into an aqueous solution of graphene oxide and noble metal salts ($HAuCl_4$, K_2PtCl_4, and Na_2PdCl_4) through the galvanic displacement of copper for the electroless deposition of noble metal nanoparticles onto silk fibroin thin films. The thermodynamic reduction potential of the underlying supporting substrate is the determining factor in governing which noble metal ions can be reduced to nanoparticles. Therefore, this approach is a platform to produce conductive silk fibroin films in a scalable, cost-efficient, rapid, and tunable process for novel biosensor applications.

References:

1. C. Bosch-Navarro, J. P. Rourkeb and N. R. Wilsona *RSC Adv*. **2016**, 73790.
2. R. S. Sundaram, C. Gomez-Navarro, K. Balasubramanian, M. Burghard and K. Kern, *Adv. Mater.* **2008**, *20*, 3050.
3. 3 X.-W. Liu, J.-J. Mao, P.-D. Liu and X.-W. Wei, *Carbon* **2011**, *49*, 477.
4. Z. Li, P. Zhang, K. Wang, Z. Xu, J. Wei, L. Fan, D. Wua and H. Zhu, *J. Mater. Chem.* **2011**, 21, 13241.
5. S. Vadahanambi, J.-H. Jung and I.-K. Oh, *Carbon* **2011**, *49*, 4449.
6. L. Qu, L. Dai, *J. Am. Chem Soc.* **2005**, *127*, 10806.
7. L. Qu, L. Dai, and E. Osawa *J. Am. Chem Soc.* **2006**, *128*, 5523.

577262 Formulation of Super Paramagnetic Iron Oxide Nanoparticles for Cryoprotecting Agent Solutions

Advances in transplantation techniques contribute significantly to improving the quality of life of patients who are affected by terminal organ failure. Despite advances in organ preservation technologies, most organs are not used because they exceed their preservation time. One technology that could enable biobanking of whole organs is cryopreservation by vitrification, where the organ is cooled and stored at a cryogenic temperature using cryoprotecting agent (CPA) solutions. Although vitrifying small tissues has proven successful, it has been difficult to preserve the structural integrity and functional physiology of bulk multicellular tissues and organs. The main challenge is associated to the current rewarming methods, where whole organs are immersed in a temperature-controlled bath, resulting in thermomechanical stresses caused by temperature gradients. One novel approach to overcoming this challenge is nanowarming using mCPA, CPA solution containing superparamagnetic iron oxide nanoparticles (SPIONs). In response to an applied alternating magnetic field (AMF) the SPIONS release heat into their environment. Using mCPA to perfuse the entire organ it will result in more uniform temperature distribution during rewarming.

Nanowarming is at a very early stage of development. Importantly, mCPA solutions containing nanoparticles that are stable for prolonged periods of time and during vitrification/nanowarming have not been reported. We address these limitations by formulating mCPAs consisting of SPIONs with high energy dissipation rates and that are stable against aggregation in CPA solutions during vitrification and nanowarming. As proof of concept, an mCPA was formulated using VS55 solution as the base CPA and hearts as the organ of interest. Colloidal stability in mCPA solutions is achieved by coating the nanoparticles with dense brushes of poly(ethylene glycol) (PEG) polymer, which confers stability in biological media. Furthermore, preliminary results demonstrate ultra-stable SPIONs in CPA pre- and post-nanowarming while achieving high temperature rise rates of up to 320oC/min at 44.4 kA/m, 276 kHz. The temperature rise is a controllable phenomenon by altering the field amplitude and/or changing the concentration of the nanoparticles in the CPA. The novel biomedical imaging technology called Magnetic Particle Imaging (MPI) was used to unambiguously and quantitatively assess the distribution of SPIONs after heart perfusion using mCPA before vitrification and after removing mCPA after rewarming. Together, these methods yield novel mCPAs suitable for perfusion, vitrification, and nanowarming of whole organs with minimal residual iron in tissues post-rewarming.

576520 Utilizing UiO-66 Metal-Organic Frameworks (MOFs) As Pulmonary Drug Delivery Vehicles

The pulmonary route is an attractive target for drug delivery applications primarily due to the large surface area of the lung, the noninvasive administration associated with inhaled therapeutics, and the high local drug concentrations for the treatment of respiratory conditions compared to oral or systemic delivery of the same molecule. The recent advances in nanotechnology have revolutionized pulmonary drug delivery by allowing for sustained drug release, controlled deposition, and limited off-target side effects. Metal-organic frameworks (MOFs) have recently emerged as potential drug carriers due to their porous structure, high loading capacity, flexible design, and tunable physiochemical properties. UiO-66, a zirconium and terephthalic acid-based MOF from the University of Oslo series has shown remarkable physical and chemical stability. Here, we examine the use of UiO-66 as a novel pulmonary drug delivery vehicle. Through an acid-free synthesis, we fabricate crystalline UiO-66 particles with tunable defectiveness, (*i.e.* porosity) which ultimately allows for the control of drug loading and release profiles. The low density and tunable defectiveness afford unique opportunities to control lung penetration and particle deposition. Furthermore, we show that UiO-66 particles have high biocompatibility and low cytotoxicity both *in vitro* and *in vivo*. Our results indicate that UiO-66 particles are efficiently internalized by lung innate immune cells, which allows for directed modulation of the lung immune microenvironment and has potential applications in inhaled immunotherapies and nanovaccines. By tuning particle properties, we demonstrate the feasibility of UiO-66 particles as a novel platform for pulmonary drug delivery.

573980 Modeling Diblock-Arm Star Polymers for Drug Delivery Applications

As the science of polymer engineering advances, many novel materials are developed that enable new applications. Star polymers present highly tunable and this flexibility has made them useful in fields such as materials science, nanoelectronics, and drug delivery. Much of their potential as nanocarriers arise from the possibility of using them as a biodegradable drug delivery system. However, in order to optimize the design of these macromolecular nanoparticles, their complex structure-property relationships and behavior in crowded environments must be well understood. To this end, molecular dynamics simulations were used to verify that diblock-arm star polymers (SPs) can be accurately described using coarse-grained models. Each arm of the star polymers examined consist of an inner hydrophobic block and an outer hydrophilic block. We observe effects of various polymer chemistries and architectures (upwards of 50 distinct SP configurations) on structure-property relationships such SP size and shape in solution, interfacial effects, relaxation times, and aggregation numbers. Results show good agreement with atomistic simulations where available and demonstrate that general trends and behaviors can be recovered. Comparisons are also made with existing theory for a subset of SP architectures and reveal the limitations of the current theoretical understanding of these materials. The ability to model these chemical species at this level will allow for simulations of polymeric nanocarriers in more complex environments including at increased concentrations and near surfaces. This work demonstrates an important first step in work that will enable improved design of polymeric nanoparticles for drug delivery applications.

576688 Improving Sustainability of CNT/Polymer Composites: Tracking CNTs Release

Carbon nanotubes (CNTs) exhibit high thermal and electrical conductivity, stiffness, tensile strength and light weight and as a result, CNTs have been employed as additives/nanofillers to enhance the properties of polymeric materials. CNTs containing polymer composites have been used in many industries including energy, environment, personal care, health care and life sciences, etc. Several studies have shown that the polymer matrix degradation can occur as a result of environmental factors such as ultraviolet (UV) radiation, moisture, mechanical stress, and/or a combination thereof. Such degradation can lead to surface accumulation, protrusion and potential release of CNTs thus it is essential to develop method and study CNTs release from composite. In this study, we investigated the effect of environmental factors on the degradation of MWCNTs/epoxy nanocomposites. Chemical changes resulting from epoxy matrix degradation were investigated by attenuated total reflectance infrared spectroscopy (ATR-FTIR) and morphological changes of the samples were assessed by scanning electron microscopy (SEM). In order to determine the rate of CNT release, we incorporated ^{14}C-labelled MWCNTs into the nanocomposites and measured their release with liquid scintillation counting (LSC) after weathering treatment. Degradation of epoxy resulted in matrix removal, exposure of CNT network, and release of a detectable amount of CNTs to the environment.

576994 Green Solvent for Conjugation of Iron Nanoparticles (Fe_3O_4) and Graphene Oxide (GO) Nanosheets to Remove Water Contaminants

Deep Eutectic Solvents (DES) have emerged as green solvents for functionalizing nanomaterials e.g., graphene oxide (GO) for introducing various functional groups on the surface. In this study, we show, for the first time, the applicability of DES as an agent for conjugating metal nanoparticles onto carbon nanomaterials to prepare nanohybrids. We conjugated Fe_3O_4 nanoparticles and GO nanosheets to prepare functionalized magnetic nanohybrids tailored for effective contaminant removal from water. The effect of the GO:Fe_3O_4 ratio in the synthesized nanohybrids (GO-Fe_3O_4) was studied at three different mass ratios of 1:1, 1:2, and 1:5 of GO to Fe_3O_4. HRTEM, XRD, Raman, and FTIR analyses were conducted to provide a better understanding of the conjugation mechanism and the physical and chemical properties of the synthesized nanohybrids. The performance of GO, Fe_3O_4, and GO-Fe_3O_4 nanohybrids as an adsorbent for removing methylene blue and lead (Pb) from aqueous solutions have been studied. Results indicated that the used green solvent i.e., DES not only can successfully conjugate Fe_3O_4 nanoparticles and the GO nanosheets, but also can provide nanohybrids with high capacity for contaminant removal from water.

575290 Continuous in Situ Sensing of Marine Pollutants Using Surface Enhanced Raman Spectroscopy (SERS)

Naturally occurring low concentrations of nitrate and phosphate in water bodies play an essential role in the nutrient cycle. At excess levels caused by human activity, however, damaging effects such as algae blooms, ecosystem disruption, and human health risks can develop. In addition, economic factors can be impacted as commercial activity is hampered. To avoid these negative effects, early detection of nitrate and phosphate concentration changes is essential. Early *in situ* detection of over-nutrification is required for rapid response and mitigation plans. Commercial nitrate and phosphate sensors utilize ultraviolet-visible (UV-vis) spectroscopy as well as various colorimetric methods. Current detection methods are either incapable of measuring at the detection limits set by the Environmental Protection Agency (EPA), or require the handling of potentially harmful chemical reagents on a research vessel. Thus, there is an urgent need for new approaches for *in situ* monitoring. Surface enhanced Raman spectroscopy (SERS) is a technique theoretically capable of single molecule detection. Unlike UV-Vis spectroscopy, SERS shows minimal interference with water, and therefore is a promising approach for nitrate and phosphate detection in coastal environments. Recent developments have led to the availability of compact and portable Raman spectrometers that could overcome the size constraints of research laboratories on marine vessels. These devices could even fit in buoys and floaters to enable real time *in situ* monitoring. However, there are clear challenges as SERS sensing is negatively affected by interference in complex media and reduced accuracy in solution. It is because of these challenges, in part, why much of the data reported in the literature are taken for purified samples that are then dried on a SERS substrate. Our goal is to address the engineering challenges for a SERS *in situ* seawater nutrient concentration measurement system. We designed and tested batch and flow-through devices that incorporate commercially available, nanostructured gold SERS substrates. By benchmarking against 4-nitrobenzenethiol/ethanol solutions as well as nitrate and phosphate/water solutions and conducting long-term studies in water, our results show that SERS devices can be used as a development platform for a seawater nutrient sensor. They also depict the challenges associated with complex media and provide insight into surface modification strategies needed to selectively detect nitrate and phosphate. This work provides a framework for the use of surface-modified nanostructured gold substrates for detection and monitoring of nitrate and phosphate in the marine environment.

575390 Electrocatalysts for the Simultaneous Removal of Nitrate and Ammonia from Wastewater

Nitrogen contamination of water resources is considered an environmental and human health issue that causes serious problems e.g., algae bloom, ecological harmful effects like food web changes, and fatal illnesses especially for newborn infants under 6 months [1]. Electro-reduction of NO_3^- and electro-oxidation of NH_3 have attracted lots of attention because they are considered a the straightforward, energy-efficient, environmentally-friendly, and easy to control techniques for NO_3^-/NH_3 detoxification [2].

Electrochemical methods for separate removal of NO_3^- and NH_3 has been extensively investigated [1, 3-5]. However, in this work, we demonstrate a new process that addresses both water pollutants simultaneously in an undivided flow cell that leads to a remarkable energy saving in the order of 35-65 % and 90 % less energy per grams of NO_3^- and NH_3 removal, respectively. A new catalyst composition material is introduced to perform both NO_3^- /NH_3 removal simultaneously. The concentration of NO_3^- and NH_3 were measured according to the Environmental Protection Agency (EPA) standards, i.e., an UV-vis spectrophotometric based technique [6], and NH_3 Ion Selective Electrode (ISE-THERMO), respectively. Effect of catalyst loading and other process operational conditions on the NO_3^- and NH_3 removal rates were investigated. Findings of this work will be presented at the meeting.

References

[1] A. Estejab, and G.G. Botte, "Ammonia Oxidation Kinetics on Bimetallic Clusters of Platinum and iridium: A Theoretical Approach" , Molecular Catalysis, vol. 445, pp. 279–292, February 2018.

[2] M. Li, Ch. Feng, Z. Zhang, Z. Shen, and N. Sugiura, "Electrochemical Reduction of Nitrate Using Various Anodes and a Cu/Zn Cathode", Electrochemistry Communications, vol. 11, pp. 1853–1856, October 2009.

[3] C. Wanag, S. Chang, M. Ye, and Q. Ren, "Current Efficiency and Energy Consumption of Electrochemical Oxidation for Ammonia Removal from Coking Wastewater using Boron-doped Diamond Electrodes", Applied Mechanics and Materials, vol. 298, pp. 1327-1332, January 2013.

[4] H. Cheng, K. Scott, and P. A. Christensen, "Paired Electrolysis in a Solid Polymer Electrolyte reactor—Simultaneously Reduction of Nitrate and Oxidation of Ammonia", Chemical Engineering Journal, vol. 108, pp. 257–268, April 2005.

[5] M. Majlesi, S. Mohseny, M. Sardar, S. Golmohammadi, and A. Sheikhmohammadi, "Improvement of Aqueous Nitrate Removal by Using Continuous Electrocoagulation/Electroflotation Unit With Vertical Monopolar Electrodes", Sustainable Environment Research, vol. 26, pp. 287-290, November 2016.

[6] T.A. Doane, and W.R. Horwath, "Spectrophotometric Detection of Nitrate with a Single Reagent", Analytical Letters, vol. 36, pp. 2713-2722, January 2003.

554336 Biogenic Metallic Nanoparticles. from Micro-Biofactories to a Nanometric Trojan Horse Approach

Antimicrobial resistance to antibiotics (AMR) and cancer and two of the main concerns that the healthcare system should face nowadays. Current drugs and antibiotic treatments are becoming ineffective or have plenty of drawbacks related to misuse and overuse. Therefore, new alternatives are needed, and nanotechnology is rising as a powerful solution over time. How the nanomaterials are created has plenty of influence in their features and applications. Traditional synthesis of nanomaterials, taking knowledge from both physics and chemistry, is subjected to several disadvantages, such as the production of toxic-by-products and harsh conditions, as well as biocompatibility issues. Green nanotechnology is presented as a suitable answer, allowing the generation of nanostructures in a cost-effective and environmentally-friendly approach employing living organisms, such as bacteria, and biomolecules.

In this research, pathogenic bacteria (both antibiotic-resistant and standard strains of Gram-negative -such as Escherichia coli- and Gram-positive -such as Staphylococcus aureus-) and human cells (both cancerous -such as human melanoma and glioblastoma cells, and healthy cell lines, such as astrocytes of fibroblasts-) were used for the synthesis of different metallic -gold (Au), palladium (Pd), platinum (Pt), gold-palladium (AuPd) and gold-platinum (AuPt)- and metalloid -selenium (Se)- nanoparticles with sizes between 5 and 120 nm surrounding by an organic-derived coating. Bacteria and human cells were cultured in the presence of metallic salts under standard conditions, allowing the generation of nanoparticles through natural detoxification processes in a synthetic protocol that is followed using microscopy and spectrophotometric techniques.

After generation, the entities are purified and extensively characterized in terms of composition, morphology and surface chemistry, through techniques such as Transmission electron microscopy (TEM), Scanning electron microscopy (SEM), Fourier-transform infrared spectroscopy (FTIR), X-Ray Photoelectron spectroscopy (XPS), X-ray powder diffraction (XRD) and Raman spectroscopy, in order to elucidate the complex organic-derived coating surrounding the nanoparticles and the metallic/metalloid core. Nanoparticles were then used as antimicrobial and anticancer agents, with effects characterized through growth curve analysis, MTS experiments and colony counting growing assays, not showing significant cytotoxicity towards healthy human cells. In order to understand the mechanisms or action, Reactive Oxygen Species (ROS) analysis was accomplished as well.

The extensive characterization of the nanoparticles showed the presence of organic molecules -such as proteins and lipids- coming from the living organisms, with selenium nanoparticles with sized between 50 and 120 nm. Bacterial tests showed an unusual selective behavior in the antimicrobial effect of bacteria-mediated nanoparticles, showing a dose-dependent inhibition when a bacterial strain X was treated with nanoparticles made by the same bacteria, while poor inhibition was found when different bacteria were used as a target. The nanoparticles showed a robust anticancer effect towards human melanoma and glioblastoma cells while remaining biocompatible towards the healthy cell lines.

On the other hand, noble mono- and bimetallic nanoparticles synthesized by human cells, with sized between 5 and 20 nm, were able to inhibit the growth of cell lines in a similar way that nanoparticles made by bacteria showed, with a certain degree of selectivity, remaining biocompatible. Besides, the production of nanomaterials induced a transition of the cells into a named zombie stage" or suspension toward the cells was not responding to chemical, physical or temporal degradation.

Therefore, we demonstrated that microbiological agents are successfully used as a synthetic machines for the generation of different metallic/metalloid nanoparticles of different compositions with biomedical properties. Therefore, they are presented as a suitable approach for the synthesis of nanomaterials in a green fashion, overcoming the limitations of traditional nanotechnology, and opening a new field for drug delivery and smart targeting of cancer and antibiotic-resistant infections.

577339 Invited Talk: Rapid Prototyping of Biohybrid Nanostructures for Sensing and Responsive Materials

Enzymes, often thought of as only biological catalysts, are also exquisitely evolved, nanoscale logic gates. They can be regulated under single or multiple inputs and have a variety of selective outputs such as bond breaking, bond formation, and shape change. These can be advantageous for many sensing and responsive material applications. Interfacing with enzymes, the (bio)-logic of nature, for such advanced materials is one mission of my group. In this talk I will present our efforts to rapidly prototype enzymes and other proteins using scalable and robust cell free protein synthesis techniques[1, 2]. I will also cover methods of interfacing with hydrolytic enzymes in closed environments, namely optical transducers based on fluorescent single walled carbon nanotubes[3] and larger radio frequency based resonant structures[4]. Applications for these biohybrid nanostructures in agricultural and medical systems will be presented such as monitoring soil health and measuring hydrolytic activity in tumor micro-environments.

1. Dopp JL, Reuel NF (2018) Process optimization for scalable E. coli extract preparation for cell-free protein synthesis. *Biochemical Engineering Journal*, 138:21–28. https://doi.org/10.1016/j.bej.2018.06.021
2. Dopp JL, Rothstein SM, Mansell TJ, Reuel NF (2019) Rapid prototyping of proteins: Mail order gene fragments to assayable proteins within 24 hours. *Biotechnology and Bioengineering*, 116(3):667–676. https://doi.org/10.1002/bit.26912
3. Kallmyer NE, Musielewicz J, Sutter J, Reuel NF (2018) Substrate-Wrapped, Single-Walled Carbon Nanotube Probes for Hydrolytic Enzyme Characterization. *Analytical Chemistry*, 90(8):5209–5216. https://doi.org/10.1021/acs.analchem.7b05444
4. Charkhabi S, Beierle AM, McDaniel MD, Reuel NF (2018) Resonant Sensors for Low-Cost, Contact-Free Measurement of Hydrolytic Enzyme Activity in Closed Systems. *ACS Sensors*, 3(8):1489–1498. https://doi.org/10.1021/acssensors.8b00267

589077 Self-Assembled Multifunctional Hybrid Nanogels for Morphogen Delivery to Stem Cells

Translation of nanoscale discoveries from the laboratory to the clinic promises new diagnostic tools, drug targeting modalities, gene therapy platforms, and tissue constructs for patients. Tissue morphogenesis during fetal development is highly dependent on spatial and temporal expression of multiple morphogens targeting different progenitor cells. During fetal development, mesenchymal stem cells (MSCs) and endothelial colony forming cells (ECFCs) are implicated in bone formation coupled by paracrine signalling between the two progenitor cells. The invading ECFCs secrete osteogenic morphogens (BMP2) to stimulate cell differentiation and mineralization whereas the differentiating MSCs release vasculogenic morphogens (VEGF) to further stimulate capillary formation for the metabolically active osteoblasts. I will present in my seminar novel self-assembled hybrid, multi-functional, self-assembled, peptide-polymer nanogels for on-demand release BMP2 and VEGF micropatterned matrices to stimulate paracrine signalling and couple osteogenesis and vasculogenesis.

572645 Plasmon Induced Photocurrent of Photosystem I Assembled on Metal Nanostructures

Photosystem I (PS I), the photosynthetic membrane protein, undergoes light activated charge separation and unidirectional electron transfer with near-unity quantum efficiency. The robust photoelectrochemical activities of PSI make it an ideal biomaterial for bio-hybrid photovoltaic and/or, optoelectronic devices. But, the first step towards rational design of such devices requires systematic electrochemical characterizations of PSI assembly in tailored biotic-abiotic interfaces. In the past, such interfaces have been created using plasmonic metal nanostructures to tune optoelectronic properties of single molecule fluorophores. Based on our recent works, we present the first-ever experimental verification of plasmon-induced photocurrent enhancements of multi-chromophore PSI complex. The excitation wavelength depended plasmon enhanced photocurrents indicate that blue light excitation is more efficient than red light excitation which can be explained by previously reported observations of excessive plasmon-induced fluorescence emission losses from PSI in red region of the excitation wavelengths. Based on these results, our on-going efforts are directed towards the systematic investigation of the effect of varying plasmon peak positions tuned with designer nanopatterns on the plasmon-enhanced photocurrents from PSI assembly on these structures. We study the role of tailored Al nanopatterned substrates designed with E-beam lithography for specific peak plasmonic resonance peaks and plasmon-induced photocurrent action spectra from PSI complexes. The FDTD simulations for localized plasmon resonance electric field distributions along with experimental results will be utilized for determining accurate plasmon enhancements action spectrum. This work will shine light on the fundamental biophysics behind the alterations in excitation energy transfer mechanism among chlorophyll network in PSI under plasmon-induced localized electric field.

574409 Effect of Di(2-ethylhexyl) Phthalate and Mono(2-ethylhexyl) Phthalate Interactions with a Lipid Membrane

Di(2-ethylhexyl) phthalate (DEHP), a common plasticizer, is present in many household products posing potential environmental exposure issues as it easily leaches from surfaces. DEHP most commonly metabolizes in the body in the presence of lipases into monoesters, such as mono(2-ethylhexyl) phthalate (MEHP). While previous studies have focused on its role as an endocrine disruptor, with major concerns of exposure to newborns and those undergoing medical procedures, DEHP and MEHP have yet to be studied for their physical and chemical interaction with lipid bilayers. These studies would provide critical information on phthalate adsorption mechanisms, potential pore formations, and compromises to the membrane structure. Previous work has investigated nanoparticle, peptide, polymer, and small molecule interactions with lipid bilayers for understanding possible toxic effects of these substances to human cells. In this work, we examined the effects of DEHP and MEHP on adsorption and permeability within a model lipid bilayer.

In our study we investigated changes occurring with vesicle and lipid bilayer structures due to the presence of various concentrations of DEHP and MEHP. We observed statistically significant changes in the size and polydispersity of L-α-phosphatidylcholine (egg PC) vesicles with the presence of 100 and 200 μM DEHP using dynamic light scattering. We then developed egg PC supported lipid bilayers (SLBs) on silica surfaces using quartz crystal microbalance with dissipation (QCM-D) monitoring. Based on frequency and dissipation changes occurring on the lipid bilayer surface, we observed mass removal or adsorption due to the presence of phthalate at concentrations of 5-200 μM, based on physiologically relevant values. With DEHP incubation, we observed lipid removal at 5 and 100 μM DEHP concentrations and significant adsorption of DEHP with 50 and 200 μM concentrations. With MEHP, we observed relatively identical lipid removal for all concentrations tested. From dynamic light scattering results of DEHP and MEHP aggregates, we hypothesize that the size of the phthalate aggregate can influence its interaction with a SLB, with larger aggregates causing adsorption to the SLB and smaller aggregates causing lipid removal. Following QCM-D studies, we performed parallel artificial permeability assay (PAMPA) by forming a 1,2-dioleoyl-sn-glycero-3-phosphocholine (DOPC) lipid bilayer across a polyvinylidene fluoride permeable support. We observed permeability differences between DEHP and MEHP with this bilayer structure and observed differences in small molecule (including caffeine, carbamazepine, warfarin, furosemide, folic acid, and methotrexate) interaction with the DOPC membrane following incubation with DEHP and MEHP. Specifically, we observed more interaction of DEHP with the DOPC bilayer compared to MEHP. We also observed that DEHP incubation caused higher molecular weight and more negative logP small molecules to interact more with the bilayer. The opposite trend was observed after MEHP incubation. Finally, we investigated the influence of adsorption of 200 μM DEHP to 1 μm chitosan particles on egg PC vesicle size. Before chitosan adsorption, egg PC vesicle size was significantly increased due to introduction and interaction with DEHP (200 μM). However, after the chitosan adsorption, there was no statistically significant change in the size of egg PC vesicles. Ultimately, this work demonstrates the physical changes DEHP and MEHP can have on lipid vesicles and bilayer structures, which can better inform exposure toxicity of phthalates.

562533 Nanoengineering of an Electroconductive Cardiac Patch

Cardiovascular disease (CVD) is the most leading cause of mortality and morbidity in the USA and costs $300 billion per year. Myocardial infarction (MI or heart attacks) is currently one of the most frequent types of CVD in the world. Among diverse technologies for rebuilding the infarcted myocardium, cardiac patches can adequately and simultaneously meet the biochemical, electrical and mechanical demands of the native heart tissue to promote regeneration following MI.

Here, we first engineered a gelatin-based porous scaffold by electrospinning technique. We then incorporated graphene nanofibers (GNFs) to the acellular porous matrices to fabricate a conductive electrospun composite scaffold. The effects of GNFs on physical properties (such as mechanical properties, degradation, swelling, surface energy, etc.), electrical conductivity and in vitro cytocompatibility of the scaffolds were evaluated. Incorporation of GNFs can bridge the electrically resistant pore walls of the matrices, to support and facilitate the internal electrical interactions between adjacent CMs. Additionally, we will align GNFs according to the electrospinning direction to produce a scaffold that can mimic the shape and orientation of CMs in natural myocardial tissue. In the next step, some epicardial-secreted factors such as cardiogenic and angiogenic factors, and proteins will be incorporated into the conductive patches obtained from the previous stage, with the aim of promoting the myocardial regeneration. It is expected that the integration of conductive GNFs and epicardial factors within 3D scaffolds may improve the therapeutic value of current cardiac patches and will open new avenues for engineering cardiac tissues.

573040 Constrained Growth of Magnetite Nanoparticles Via Artificial Biomineralization

Biomineralization is a process widely used in nature to direct the growth of inorganic structures such as bones, teeth, and shells, often leveraging the versatile nature of proteins to precisely direct the morphologies and crystal structures of these mineralized tissues. This work uses a biomimetic approach to replicate the formation of iron oxide nanoparticles by ferritin, an iron storage protein found in nearly all organisms, using a much simpler peptide system. Instead of a large and complex protein, we used a polyelectrolyte complex micelle, consisting of two oppositely charged polymer blocks that bind electrostatically to form a core, and a neutral hydrophilic block forming a corona; our micelle consists of PEG-b-polylysine and polyglutamic acid. When iron oxide synthesis is performed in the presence of these micelles, size-constrained Fe_3O_4 particles are produced and suspended in aqueous solution. Dynamic light scattering, electron microscopy, and small-angle x-ray scattering results confirm that the average particle diameter is below 100 nm and that the size distribution is monodisperse, which suggests that the oxide nanoparticles and peptides form a hybrid structure with the nanoparticle contained within the micelle. Furthermore, the particle morphology and stability in suspension can be tuned by adjusting parameters such as the concentration of reactants or type of polymers present (polyelectrolyte micelles or polyglutamic acid alone). Due to the magnetic properties of iron oxide, these particles could have diverse applications in memory storage devices, spintronics, and nanomedicine.

AUTHOR INDEX

AUTHOR INDEX

AUTHOR INDEX

AUTHOR INDEX

AUTHOR INDEX

AUTHOR INDEX

AUTHOR INDEX

AIChE
120 Wall Street, FL 23
New York, NY 10005-4020

ISBN 978-1-7138-0543-4

9 781713 805434

CURRAN ASSOCIATES INC.
proceedings.com